# Student Solutions Manual

for

Moore/Notz/Fligner's

# The Basic Practice of Statistics

*Sixth Edition*

R. Scott Linder
*Ohio Wesleyan University*

W. H. Freeman and Company
New York

©2013 by W.H. Freeman and Company

ISBN-13: 978-1-4292-8000-6
ISBN-10: 1-4292-8000-X

All rights reserved.

Printed in the United States of America

First printing

W.H. Freeman and Company
41 Madison Avenue
New York, NY 10010
Houndmills, Basingstoke RG21 6XS England

www.whfreeman.com

# Contents

| | |
|---|---|
| Chapter 1: Picturing Distributions with Graphs | 1 |
| Chapter 2: Describing Distributions with Numbers | 8 |
| Chapter 3: The Normal Distributions | 14 |
| Chapter 4: Scatterplots and Correlation | 19 |
| Chapter 5: Regression | 27 |
| Chapter 6: Two-Way Tables* | 39 |
| Chapter 7: Exploring Data: Part I Review | 43 |
| Chapter 8: Producing Data: Sampling | 49 |
| Chapter 9: Producing Data: Experiments | 52 |
| Chapter 10: Introducing Probability | 57 |
| Chapter 11: Sampling Distributions | 61 |
| Chapter 12: General Rules of Probability* | 65 |
| Chapter 13: Binomial Distributions* | 71 |
| Chapter 14: Confidence Intervals: The Basics | 75 |
| Chapter 15: Tests of Significance: The Basics | 79 |
| Chapter 16: Inference in Practice | 83 |
| Chapter 17: From Exploration to Inference: Part II Review | 87 |
| Chapter 18: Inference about a Population Mean | 91 |
| Chapter 19: Two-Sample Problems | 98 |
| Chapter 20: Inference about a Population Proportion | 105 |
| Chapter 21: Comparing Two Proportions | 110 |
| Chapter 22: Inference about Variables: Part III Review | 116 |
| Chapter 23: Two Categorical Variables: The Chi-Square Test | 120 |
| Chapter 24: Inference for Regression | 129 |
| Chapter 25: One-Way Analysis of Variance: Comparing Several Means | 139 |

# Chapter 1: Picturing Distributions with Graphs

1.1: (a) The individuals are the car makes and models. (b) For each individual, the variables recorded are Vehicle type (categorical), Transmission type (categorical), Number of cylinders (usually treated as quantitative), City mpg (quantitative), Highway mpg (quantitative), and Carbon footprint (tons, quantitative).

1.3: (a) If we sum the shares given, they sum to 67.3%. Since these don't sum to 100%, there must be other formats. Hence, 100% − 67.3% = 32.7% of the radio audience listens to stations with other formats. (b) a bar graph is provided. Note that "O" is used to denote the 32.7% of listeners in other formats, and this group is the largest.

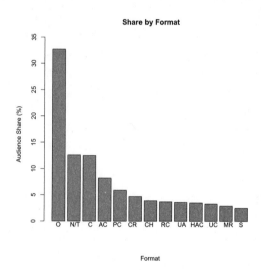

(c) A pie chart would be inappropriate based only on the data presented because the areas of the pie wedges would be relative to the total of the categories presented (67.3%). For example, consider the "Country" format, which captures 12.5% of the audience. Since the total of the shares provided is 67.3%, the area of the wedge corresponding to "Country" would be 12.5/67.3, or 18.6% of the total pie. This is deceptive, since the Country format accounts for only 12.5% of the audience. If we include a wedge for "other" that accounts for 32.7% of the total, a pie chart would be reasonable.

1.5: A pie chart would make it more difficult to distinguish between the weekend days and the weekdays. Many medical professionals may work regular Monday through Friday hours. Hence, some births are scheduled (induced labor, for example), and probably most are scheduled for weekdays.

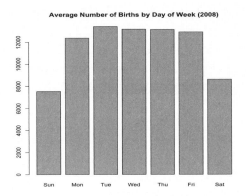

1.7: Use the applet *One Variable Statistical Calculator* to answer these questions.

1.9: (a) We see that the District of Columbia has an outlying percentage of never-married women. This isn't surprising because the District of Columbia is the center of Federal government and hence has an enormous number of young professionals, many of whom may not be married. We would also expect the percentage of never-married men to be high here also. (b) The 26th ordered value falls between 26 and 28 percent. The values in this distribution fall between 20 and 54, but virtually all are between 20 and 34. Again, the District of Columbia is an outlier.

1.11: Here is a stemplot for health expenditure per capita (PPP). Data are rounded to units of hundreds. For example, Argentina's "1332" becomes 13. Stems are thousands, and are split, as prescribed.

```
0  1 1 2 3
0  7 7 7 8 8 8 8 8
1  0 3
1  7
2  3
2  7 7 7 7 8
3  0 3 3 4 4
3  5 5 6 7 8 9
4  4
4  8
5
5
6
6
7  3
```

# Solutions

This distribution is somewhat right-skewed, with a single high outlier (United States). There are two clusters of countries. The center of this distribution is around 25 ($2500 spent per capita), ignoring the outlier. The distribution's spread is from 1 ($100 spent per capita) to 73 ($7300 spent per capita).

1.13: (a) the students.

1.15: (b) Square footage and average monthly gas bill are both quantitative variables.

1.17: (b) 20% to 22%.

1.19: (c) 30.9 minutes.

1.21: (b) close to 23.4 minutes. Take the 26th ordered value.

1.23: (a) Individuals are students who have finished medical school. (b) 6, including "Name." "Age" and "USMLE" are quantitative. The others are categorical.

1.25: "Other colors" should account for 4%, since the indicated colors account for a total of 96%. A bar graph would be an appropriate display, since the data are categorical:

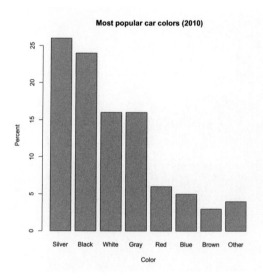

1.27: (a) A bar graph follows. (b) To make a pie chart, you would need to know the total number of deaths in this age group, or (equivalently) the number of deaths due to "other" causes.

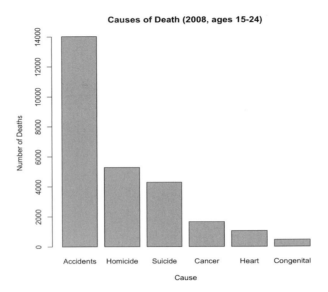

1.29: (a) A bar graph is provided below. (b) A pie chat would be inappropriate, because these percentages aren't "shares." That is, the percentages don't sum to 100%.

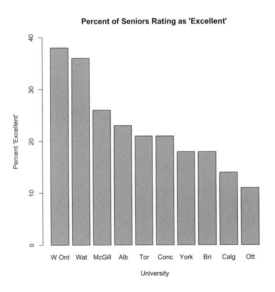

1.31: (a) Ignoring the four lower outliers, the distribution is roughly symmetric, centered at a score of about 110, and having spread in scores of 86 to 136. (b) 64 of the 78 scores are more than 100. This is 82.1%.

1.33:

1. Are you male or female → Histogram (c). There are two outcomes possible, and the difference in frequencies is likely to be smaller than the right-handed/left-handed difference in (2).

2. Are you right-handed or left-handed → Histogram (b), since there are more right-handed people than left handed people, and the difference is likely larger than the sex difference in (1).

3. Heights → Histogram (d). Height distribution is likely to be symmetric.

4. Time spent studying → Histogram (a). The variable takes on more than one value, and time spent studying may well be a right-skewed distribution, with most students spending less time studying, and some students spending more time studying.

1.35:

(a) States vary in population, so you would expect more nurses in California than in New Hampshire, for example. Nurses per 100,000 provides a better measure of how many nurses are available to serve a state's population. That is, we should compare Nurse densities by population, adjusted for the size of the population of a state. (b) A histogram is provided below, where we compare states by Nurses per 100,000 population. The District of Columbia, South Dakota, and Massachusetts are the three states different from the others. Perhaps they could be considered outliers. It's difficult to know why these states would have more nurses than other states.

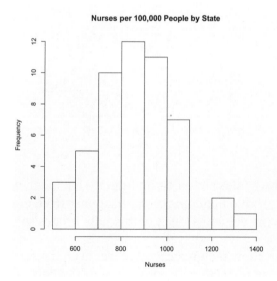

**1.37:** Here is a stemplot for the pups data, using split stems at the tens place. This is a right-skewed distribution, with center around 25 pups and spread of 17 pups to 56 pups. There were several extremely good years for pups, resulting in more than 45 births.

```
1  777789
2  0122344
2  555579
3  12333
3  899
4  3
4  77
5  4
5  6
```

**1.39:** A time plot of seal pups. The decline in population is not described by the stemplot made in Exercise 1.37.

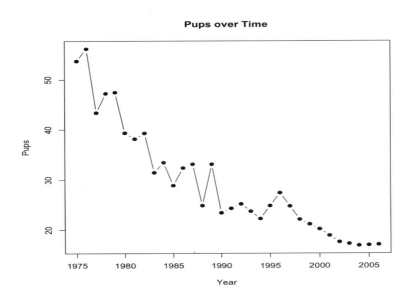

**1.41:** Coins with earlier (lower) dates are older, and rarer. The number of coins manufactured each year varies substantially, but older coins often disappear from circulation. Hence, there are more coins with larger dates (newer coins) than with smaller dates (older coins).

**1.43:** (a) Graph (a) appears to show the greatest increase, even though both plots describe the same data. Vertical scaling can impact one's perception of the data. (b.) In both graphs, tuition starts around $2000 and rises to $7700. Again, both plots describe the same data.

**1.45:** (a) A time plot of ozone hole size (area) is provided below. There is a trend, as well as year-to-year variability. The hole has grown a lot over the period studied, but may have leveled out in recent years. Notice that there is a good deal of year-to-year variability in the measured ozone hole size. This could be because the hole varies in size substantially from year to year, or it could be due to variation in the measuring device(s) or method(s) used to measure the hole. This is why it's important to collect the data over time – this plot makes fairly clear that the size of the hole increased substantially over the time measured. However, it is difficult to know whether the "flat" part of the graph in recent years is due to a leveling off, or is an artifact of the variation.

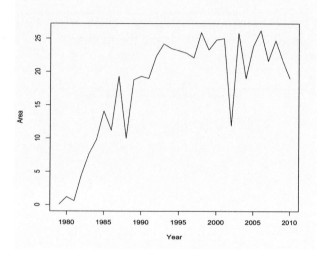

(b) A stemplot of ozone hole size (area) is provided below. The midpoint is 19.3 millions of km$^2$. A stemplot fails to capture the relationship between size of hole and year, simply because the plot uses only one variable and ignores year.

| 0 | 0004 |
|---|---|
| 0 | 79 |
| 1 | 0114 |
| 1 | 899999 |
| 2 | 11222333444 |
| 2 | 5556 |

# Chapter 2: Describing Distributions with Numbers

2.1: Mean breaking strength = 30841 pounds. Only 6 pieces have strengths less than the mean. The mean is so small relative to the data because of the sharp left skew (low outliers).

2.3: The mean travel time is 31.25 minutes. The median travel time is 22.5 minutes. The mean is significantly larger than the median due to the right skew in the distribution of times.

2.5: A histogram is given below. There is a strong right skew, so we anticipate that the mean is larger than the median. Indeed, the mean is 4.61 and the median is 3.95 tons per person, as computed using software.

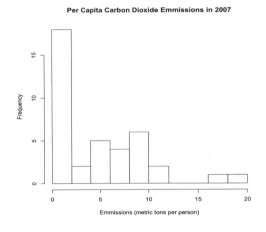

2.7: (a) Minimum = 9, Q1 = 16, Median = 18, Q3 = 22, Maximum = 51. (b) The boxplot shows right skew in the distribution of MPG values.

2.9: Using the data values obtained in Problem 2.7, the interquartile range is given by IQR = 22 − 16 = 6. Hence, the Upper Fence is Q3 + 1.5(IQR) = 22 + 1.5(6) = 31. There are five values greater than 31 that would be identified as potential outliers (33, 35, 41, 41, 51). The Lower Fence is given by Q1 − 1.5xIQR = 16 − 1.5(6) = 7. There are no potential outliers in the left tail, since no observations are below 7.

2.11: This problem illustrates that two data sets can have the same mean and standard deviation yet vary in their distribution shapes. Here, both data sets have the same mean and standard deviation (about 7.5 and 2.0, respectively). However, construct simple stemplots to reveal that Data A have a very left-skewed distribution, while Data B have a slightly right-skewed distribution, as seen in the histograms provided below.

# Solutions

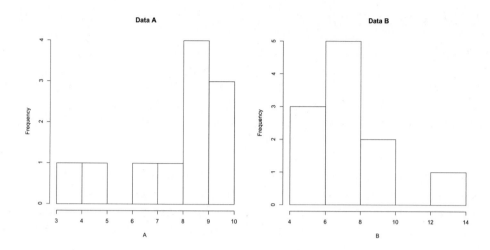

**2.13: STATE:** We compare tree counts in three types of plots: Group 1 plots were never logged; Group 2 plots were logged 1 year ago; Group 3 plots were logged 8 years ago. Is there a difference in number of trees between these plots? **PLAN:** We compute summary statistics and construct histograms for all three groups, then use them to compare. **SOLVE:** Histograms for all three groups are provided. It's important to use the same scale on the horizontal axis, so that a comparison can be made easily. The table below provides summary statistics (mean and standard deviation) for all three groups.

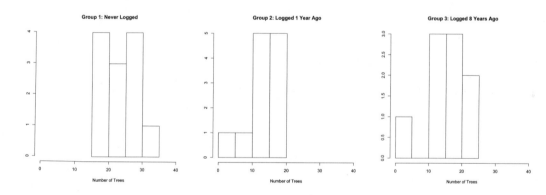

Group 1: $\bar{x} = 23.7500$, $s = 5.06548$ trees.
Group 2: $\bar{x} = 14.0833$, $s = 4.98102$ trees.
Group 3: $\bar{x} = 15.7778$, $s = 5.76146$ trees.

**CONCLUDE:** The histograms and summary statistics suggest that while trees do grow back, most of the growth occurs during the first year. Clearly it takes longer than eight years for full recovery from the impact of logging.

2.15: (b) 167.48

2.17: (b) 151.6, 163.5, 168.25, 174.3, 177.6

2.19: (b) 50%.

2.21: (c) 8.2.

2.23: (b) seconds.

2.25: Distributions such as incomes are typically very right-skewed, because the rare incomes are larger, and (relatively) lower incomes are more common. The distribution of incomes in this group is almost certainly right-skewed, so the mean is larger than the median. Hence, the mean income is $58,762 and the median income is $46,931.

2.27: We have an even number of colleges (842), so the median location is (842 + 1)/2 = 421.5. Hence, the median is computed by averaging the 421st and 422nd endowment sizes. The 1st quartile, Q1, is found by taking the median of the first 421 endowments (when sorted). This would be the (421 + 1)/2 = 211th endowment. Similarly, Q3 is found as the 632nd endowment (211 endowments above the median).

2.29: The five-number summaries for the three species are tabulated below. Box plots don't add much information not already present in the stemplots.

|  | Minimum | Q1 | Median | Q3 | Maximum |
| --- | --- | --- | --- | --- | --- |
| *Bihai* | 46.34 | 46.71 | 47.12 | 48.25 | 50.26 |
| Red | 37.4 | 38.07 | 39.16 | 41.69 | 43.09 |
| Yellow | 34.57 | 35.45 | 36.11 | 36.82 | 38.13 |

2.31: A histogram of the survival times follows. The distribution is strongly right-skewed, with center around 100 days, and spread 0 to 600 days. (b) Because of the extreme right skew, we should use the five-number summary. This is given by Minimum = 43 days, Q1 = 82.5 days, Median = 102.5 days, Q3 = 151.5 days, and Maximum = 598 days. Notice that the median is closer to Q1 than to Q3, which is consistent with a strong right-skew.

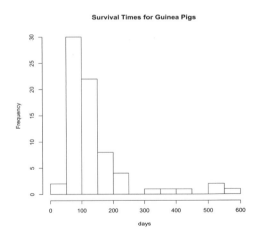

2.33: (a) Symmetric distributions. (b) Removing the outliers reduces both means and both standard deviations.

2.35: (a) The 6th observation must be placed at median for the original 5 observations. (b) No matter where you put the 7th observation, the median is one of the two repeated values above.

**2.37:** The mean for all 51 entries is 8.4%, far from the national percentage of 12.5%. You can't average averages. Some states, like California and Florida, are larger and should carry more weight in the national percentage. Indeed, there are more people over the age of 65 living in Florida than there are residents in Wyoming.

**2.39:** (a) Pick any four numbers all the same: for instance, (4,4,4,4) or (6,6,6,6). (b) (0,0,10,10). (c) There is more than one possible answer for (a), but not for (b).

**2.41:** Lots of answers are possible. Start by insuring that the median is 7, by "locking" 7 as the 3rd smallest value. Then, adjust the minimum or maximum accordingly to acquire a mean of 10 (so they sum to 50). One solution: 5 6 7 8 24.

**2.43:** (a) Weight losses that are negative correspond to weight *gains*. (b) A side-by-side boxplot (a version that reports suspected outliers using the 1.5 IQR rule) is provided below. Gastric banding seems to produce higher weight losses, typically. (c) It is better to measure weight loss relative to initial weight. (d) If the subjects that dropped out had continued, the difference between these groups would be as great or greater because many of the "lifestyle" dropouts had negative weight losses (i.e., weight gains), which would pull that group down.

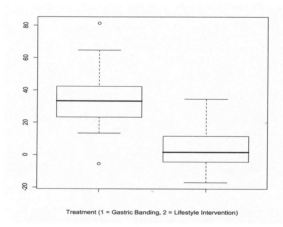

**2.45:** The distribution of average returns is skewed-left. Most years, average return is positive. Returns range from about –40% to 40%, with the median return about 16%.

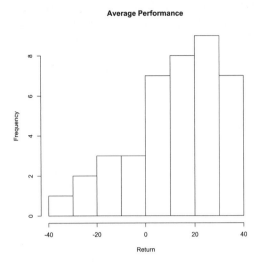

2.47: Based on side-by-side boxplots, lean people spend relatively more time active, but there is little difference in the time these groups spend lying down.

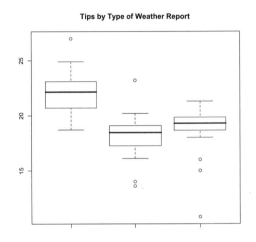

2.49: The distribution is also right-skewed. The median salary was $300, and the middle half of salaries were between $167.50 and $450. A handful of Canadians made $1000 or more. One earned $2200.

2.51: (a) Min = 0.0272, Q1 = 0.6449, Median = 3.954, Q3 = 8.1555, Max = 18.9144. Notice that the maximum is farther from Q3 than the minimum is from Q1. This suggests right skew. (b) IQR = 8.1555 − 0.6449 = 7.5106. Hence, 1.5(IQR) = 11.2659. Now Q1 − 1.5(IQR) = 0.6449 − 11.2659 < 0, so no values are more than 1.5 IQR's below Q1. Also, Q3 + 1.5(IQR) = 8.1555 + 11.2659 = 19.4214, so there are no high outliers. This rule is rather conservative — most people would easily call the United States' value (18.9144) a far outlier, and perhaps Canada would be considered an outlier, too.

2.53: In Exercise 2.49 the 1st and 3rd quartiles were given by $167 and $450, respectively. Hence, the Interquartile Range is given by $450 − $167 = $283. The Upper Fence is given by $450 + 1.5($283) = $874.50. Any of the 11 incomes more than $874.50 would be considered an outlier by the 1.5 IQR rule.

# Chapter 3: The Normal Distributions

3.1: Sketches will vary. (a) A symmetric distribution with two peaks is provided below, on the left. (b) A distribution with a single peak and skewed to the left is provided below, on the right.

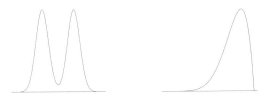

3.3: $\mu = 2.5$, which is the obvious balance point of the rectangle. The median is also 2.5 because the distribution is symmetric (so that the mean and median are the same), and half the area under the curve lies to the left and half to the right of 2.5.

3.5: Here is a sketch of the distribution of the Normal curve describing thorax lengths of fruit flies. The tick marks are placed at the mean, and at one, two and three standard deviations above and below the mean for scale. For example, thorax lengths one standard deviation above average are 0.800 *mm* + 0.078 *mm* = 0.878 *mm*. Also, thorax lengths two standard deviations below average are 0.800 *mm* – 2(0.078 *mm*) = 0.644 *mm*.

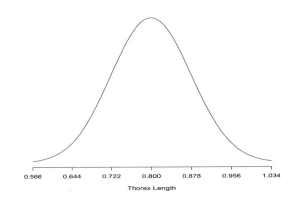

3.7: (a) In 95% of all years, monsoon rain levels are between 688 and 1016 *mm*. These values are two standard deviations above and below the mean: 852 ± 2(82) = 688 to 1016 *mm*. (b) The driest 2.5% of monsoon rainfalls are less than 688 *mm*; this is more than two standard deviations below the mean.

3.9: We need to use the same scale. Recall that 6 feet = 72 inches. A woman 6 feet tall has standardized score $z = \dfrac{72 - 64.3}{2.7} = 2.85$ (quite tall, relatively). A man 6 feet tall has standardized score $z = \dfrac{72 - 69.9}{3.1} = 0.68$. Hence, a woman 6 feet tall is 2.85 standard deviations taller than average for women. A man 6 feet tall is only 0.68 standard deviations above average for men.

3.11: Let $x$ be the monsoon rainfall in a given year. (a) $x \leq 697$ mm corresponds to $z \leq \frac{697-852}{82} = -1.89$, for which Table A gives $0.0294 = 2.94\%$. (b) $683 < x < 1022$ corresponds to $\frac{683-852}{82} < z < \frac{1022-852}{82}$, or $-2.06 < z < 2.07$. This proportion is $0.9808 - 0.0197 = 0.9611 = 96.11\%$.

3.13: (a) We want the value such that the proportion below is 0.15. Using Table A, looking for an area as close as possible to 0.1500, we find this value has $z = -1.04$ (software would give the more precise $z = -1.0364$). (b) Now we want the value such that the proportion above is 0.70. This means that we want a proportion of 0.30 below, Using Table A, looking for an area as close to 0.3000 as possible, we find this value has $z = -0.52$ (software gives $z = -0.5244$).

3.15: (b) Income distributions are typically skewed to the right, as there are more relatively low incomes than relatively high incomes. Also, in a forest, there are likely to be many more relatively short trees than there are relatively tall trees, so the distribution of 100 white pine trees will tend to be skewed right. Although the distribution of home prices in a very large metropolitan area tends to be right-skewed, perhaps in a suburb, where the houses tend to be similar, the distribution is more symmetric.

3.17: (b) The curve is centered at 2.

3.19: (b) $266 \pm 2(16) = 234$ to 298 days.

3.21: (b) $z = \frac{127-100}{15} = 1.80$.

3.23: (a) 0.2266.

3.25: Sketches will vary, but should be some variation on the one shown here: the peak at 0 should be "tall and skinny," while near 1, the curve should be "short and fat."

3.27. 70 is two standard deviations below the mean (that is, it has standard score $z = -2$), so about 2.5% (half of the outer 5%) of adults would have WAIS scores below 70.

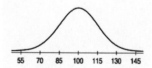

**3.29:** (a) We want the proportion less than $z$ to be 0.60, so looking up a left-tail area of 0.6000 in the table, we find $z = 0.25$. (Software gives $z = 0.2533$.) (b) If 15% are more than $z$, then 85% are less than or equal to $z$. Hence, $z = 1.04$. (Software gives $z = 1.0364$.)

**3.31:** About 0.2119: The proportion of rainy days with rainfall pH below 5.0 is about 0.2119: $x < 5.0$ corresponds to $z < \dfrac{5.0 - 5.43}{0.54} = -0.80$, for which Table A gives 0.2119.

**3.33 (Expanded):** For the $N(0.8750, 0.0012)$ distribution, $0.8720 < x < 0.8780$ corresponds to $\dfrac{0.8720 - 0.8750}{0.0012} < z < \dfrac{0.8780 - 0.8750}{0.0012}$, or $-2.50 < z < 2.50$, for which Table A gives $0.9938 - 0.0062 = 0.9876$.

For problems 3.35–3.37, let $x$ denote the gas mileage of a randomly selected vehicle type from the population of 2010 model vehicles (excluding the high mileage outliers, as mentioned).

**3.35:** Cars with better mileage than the Camaro correspond to $x > 19$, which corresponds to $z > \dfrac{19 - 20.3}{4.3} = -0.30$. Hence, this proportion is $1 - 0.3821 = 0.6179$, or 61.79%.

**3.37:** As seen in Example 3.11, the first and third quartiles have $z = -0.67$ and $z = 0.67$, respectively. Hence, the first quartile is $20.3 - (0.67)(4.3) = 17.42$ mpg, and the third quartile is $20.3 + (0.67)(4.3) = 23.18$ mpg.

**3.39:** If William scored 32, his percentile is simply the proportion of all scores lower than 32. Let $x$ be the MCAT score for a randomly selected student that took it. The event $x < 32$ corresponds to $z < \dfrac{32 - 25.0}{6.4} = 1.09$. Hence, 0.8621 is the corresponding proportion, or 86.21%. William's MCAT score is the 86.21 percentile.

**3.41:** If $x$ is the height of a randomly selected woman in this age group, we want the proportion corresponding to $x > 69.9$ inches. This corresponds to $z > \dfrac{69.9 - 64.3}{2.7} = 2.07$, which has proportion $1 - 0.9808 = 0.0192$, or 1.92%.

3.43: (a) Let $x$ be a randomly selected man's SAT math score. The event that a man's SAT math score is greater than 750 is then $x > 750$, and corresponds to $z > \frac{750-534}{118} = 1.83$. That is, a man's SAT math score of 750 is 1.83 standard deviations above average. By Table A, the proportion of scores below $z = 1.83$ is 0.9664. Hence, the proportion of men scoring higher than 750 on the SAT math test is $1 - 0.9664 = 0.0336$. (b) Let $x$ be a randomly selected woman's SAT math score. $x > 750$ corresponds to $z > \frac{750-500}{112} = 2.23$. By Table A, the proportion of scores below $z = 2.23$ is 0.9871. Hence, the proportion of women scoring above 750 on the SAT math test is $1 - 0.9871 = 0.0129$.

3.45. (a) About 0.6% of healthy young adults have osteoporosis (the cumulative probability below a standard score of $-2.5$ is 0.0062). (b) About 31% of this population of older women has osteoporosis: The BMD level that is 2.5 standard deviations below the young adult mean would standardize to $-0.5$ for these older women, and the cumulative probability for this standard score is 0.3085.

3.47: (a) $145{,}000/1{,}568{,}835 = 0.0924$, or 9.24%. (b) There are $50{,}860 + 145{,}000 = 195{,}860$ students with ACT score 28 or higher. This is $195{,}860/1{,}568{,}835 = 0.1248$, or 12.48%. (c) If $x$ is the ACT score, then $x > 28$ corresponds to $z > \frac{28-21.0}{5.2} = 1.35$, so the corresponding proportion is $1 - 0.9115 = 0.0885$, or 8.85%.

3.49: (a) A histogram is provided below, and appears to be roughly symmetric with no outliers. A Normal approximation seems reasonable. (b) Mean = 544.42, Median = 540, Standard deviation = 61.24, Q1 = 500, Q3 = 580. The mean and median are close, and the distances of each quartile to the median are equal. These results are consistent with a Normal distribution. (c) If $x$ is the score of a randomly selected GSU entering student, then we are assuming $x$ has the $N(544.42, 61.24)$ distribution. The proportion of GSU students scoring higher than the national average of 501 corresponds to the proportion of $x > 501$, or $z > \frac{501-544.42}{61.24} = -0.71$, or $1 - 0.2389 = 0.7611$, or 76.11%. (d) In fact, 1776 entering GSU students scored higher than 501, which represents $1776/2417 = 0.7348$, or 73.48%. The nominal Normal probability in (c) fits the actual data well.

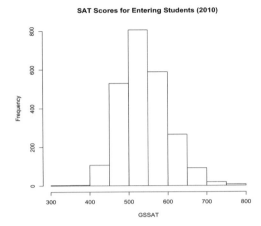

3.51: (a) The 65 Canadians with earnings greater than $375 represent 65/200 = 0.325, or 32.5%. $x > 375$ corresponds to $z > \frac{375 - 350.30}{292.20} = 0.08$, which has proportion $1 - 0.5319 = 0.4681$, or 46.81% above.  (b) $x < 0$ corresponds to $z < \frac{0 - 350.30}{292.20} = -1.20$, or 0.1151, or 11.51%.  (c) The Normal distribution model predicts 11.5% of Canadians to earn less than $0, while (of course) none do. This is a substantial error since the Normal model predicts 11.5% of values more than 375, where we actually observed 32.5% more than 375. The standard deviation ($292.20) is large relative to the average ($350.30), which suggests a strong right skew in the distribution, given that no values can be negative. In this application, the data seem to be far from Normal in distribution.

3.53: Because the quartiles of any distribution have 50% of observations between them, we seek to place the flags so that the reported area is 0.5. The closest the applet gets is an area of 0.5034, between −0.680 and 0.680. Thus the quartiles of any Normal distribution are about 0.68 standard deviations above and below the mean.

**Note**: *Table A places the quartiles at about 0.67; other statistical software gives ±0.6745.*

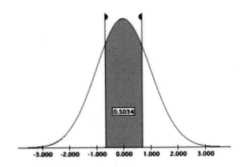

# Chapter 4: Scatterplots and Correlation

4.1: Recall that the response variable is an outcome of a study. The explanatory variable is a variable that explains or influences changes in a response variable. (a) We tend to believe that time spent studying influences the grade. Hence, time spent studying is explanatory; the grade is the response variable. (b) It' is true that taller people tend to be heavier, and heavier people tend to be taller… but here we are probably interested in exploring the relationship between the variables. Hence, there is no reason to view one or the other as explanatory. (c) We might believe that more time spent on Facebook tends to result in lower grades, to some extent. Hence, time spent online using Facebook is explanatory, GPA is the response variable. (d) It would not be surprising to learn that students that score well on one of these tests tend to score well on the other. However, we have no reason to list either exam score as an explanatory variable. Our goal here is to explore the relationship between the test scores.

4.3: For example: weight, sex, other food eaten by the students, beer type (light, imported, . . . ).

4.5: Our goal here is to determine whether outsourcing maintenance has an impact on the percentage of flights delayed. Here, Outsource percent is the explanatory variable and should be on the horizontal axis. Delay percent is the response and should be on the vertical axis. The scatterplot is provided below. Notice that the observation in the lower-right corner of the plot (Hawaiian Airlines) corresponds to 74.1% Outsource percentage and 7.94% Delay percentage. This observation is an outlier. This observation is discussed in Problem 4.7. There is, at best, a very weak relationship between Outsource percentage and Delay percentage.

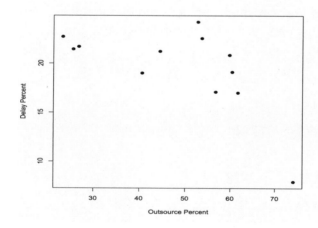

4.7: There is an outlier (Hawaiian Airlines) corresponding to 74.1% Outsource percentage and 7.94% Delay percentage. Removing it, we would see no association between these variables. Without removing it, there is, at best, a very weak, negative association between the variables. In general there seems to be little or no relationship between outsourcing of maintenance and flight delays.

4.9: (a) Women are marked with filled circles, men with open circles. (b) For both men and women, the association is linear and positive. The women's points show a stronger association. As a group, males typically have larger values for both variables (they tend to have more mass, and tend to burn more calories per day).

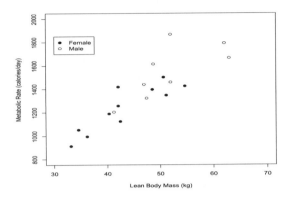

4.11: $r$ would not change; units do not affect correlation.

4.13: In computing the correlation, note that $\bar{x} = 50$ mph, $s_x = 15.8114$ mph, $\bar{y} = 26.8$ mpg and $s_y = 2.6833$ mpg. Refer to the table of standardized scores below, then note that $r = \dfrac{\sum z_x z_y}{n-1}$ = 0/4 = 0. The correlation is zero because these variables do not have a straight-line relationship; the association is neither positive nor negative. Remember that correlation only measures the strength and direction of a *linear* relationship between two variables. Note, however, that the scatterplot does indicate that there is a relationship between $X$ and $Y$... it just isn't a straight-line relationship, so correlation is zero.

| $z_x$ | $z_y$ | $z_x z_y$ |
|---|---|---|
| −1.2649 | −1.0435 | 1.3199 |
| −0.6325 | 0.4472 | −0.2828 |
| 0 | 1.1926 | 0 |
| 0.6325 | 0.4472 | 0.2828 |
| 1.2649 | −1.0435 | −1.3199 |
| | | 0 |

*Solutions*

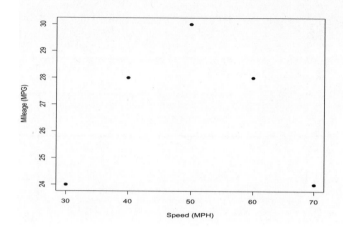

**4.15:** (a) The association should be positive (e.g., if oil prices rise, so do gas prices).

**4.17:** (a) 0.9. Without the outlier, there is a strong positive linear relationship.

**4.19:** (c) A correlation close to 0 might arise from a scatterplot with no visible pattern, but there could be a nonlinear pattern. See Exercise 4.13, for example.

**4.21:** (a) 1. There would be a perfect, positive linear association.

**4.23:** (b) Computation with calculator or software gives $r = 0.8900$.

**4.25:** (a) Overall, there is a slightly negative association between these variables. (b) There is general disagreement — low BRFSS scores correspond to greater happiness, and these are associated with higher-ranked states (the least happy states, according to the objective measure). (c) It is hard to declare any of the data values as "outliers." It does not appear that any of the values are obviously outside of the general pattern. Perhaps one value (Rank = 8, BRFSS = 0.30) is an outlier, but this is hard to say.

**4.27:** (a) The scatterplot suggests a strong positive linear association between distance and time with respect to the spread of Ebola. (b) $r = 0.9623$. This is consistent with the pattern described in (a). (c) Correlation would not change, since it does not depend on units.

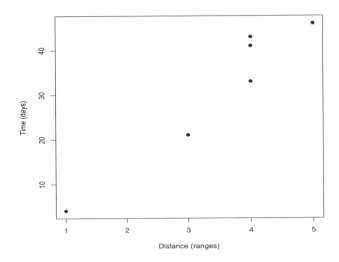

4.29: (a) The scatterplot is shown; note that neural activity is explanatory (and so should be on the horizontal axis). (b) The association is moderately strong, positive, and linear. The outlier is in the upper right corner. (c) For all points, $r = 0.8486$. Without the outlier, $r = 0.7015$. The correlation is greater with the outlier because it fits the pattern of the other points; if one drew the line suggested by the other points, the outlier would extend the length of the line and would therefore decrease the relative scatter of the points about that line.

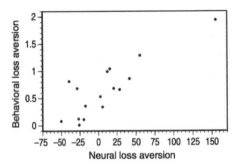

4.31: (a) The scatterplot is provided below. (b) The plot suggests that there is a strong relationship between alcohol intake and relative risk of breast cancer (again, this is an observational study, so no causal relationship is established here). It seems that type of alcohol has nothing to do with the increase since the same pattern and rate of increase is seen for both groups.

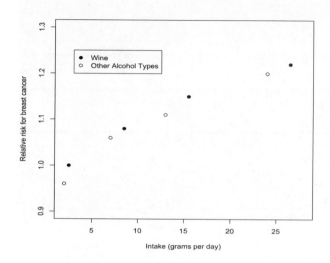

4.33: (a) A plot follows, and suggests that "Good" weather reports tend to yield higher tips. (b) The explanatory variable is categorical, not quantitative, so $r$ cannot be used. Notice that we can arrange the categories any way, and these different arrangements would suggest different associations. Hence, it doesn't make sense to discuss a relationship direction here.

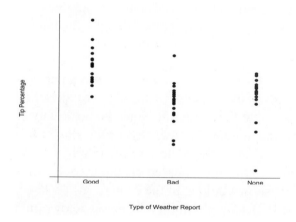

4.35: (a) The scatterplot is provided below. Changing the units has a dramatic impact on the plot. (b) Nevertheless, units do not impact correlation, as correlation is unit-free. For both data sets, $r = 0.8900$.

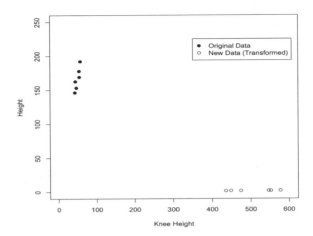

4.37: (a) Small-cap stocks have a lower correlation with municipal bonds, so the relationship is weaker. (b) She should look for a negative correlation (although this would also mean that this investment tends to *decrease* when bond prices rise).

4.39: (a) Because gender has a nominal scale, we cannot compute the correlation between sex and any other variable. There is a strong *association* between sex and income. Some writers and speakers use "correlation" as a synonym for "association," but this is not correct. (b) A correlation of $r = 1.09$ is impossible, because $r$ is restricted to be between $-1$ and $1$. (c) Correlation has no units, so $r = 0.63$ centimeter is incorrect.

4.41: (a) Because two points determine a line, there is always a perfect linear relationship between two points, and the correlation is always 1 or –1. (b) Sketches will vary; an example is shown below, upper-left. Note that the scatterplot must be positively sloped, but $r$ is affected only by the scatter about the line, not by the steepness of the slope of that line. (c) The first nine points cannot be spread from the top to the bottom of the graph because in such a case the correlation cannot exceed about 0.66 (this is based on experience — lots of playing around with the applet). One possibility is shown below, in the center. (d) To have $r = 0.7$, the curve must be higher at the right than at the left. One possibility is shown below, on the right.

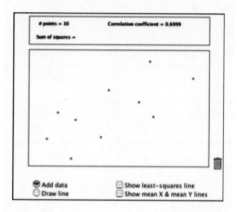

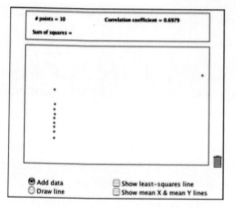

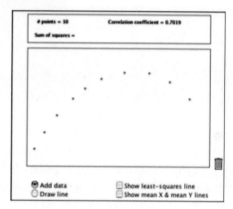

4.43: STATE: Is there a relationship between improved running times and improved running times of women? PLAN: To study the improvements in running times between men and women, we'll plot the data on the same scatterplot. We will not use correlation, but we will examine the plot to see if women are beginning to outrun men. SOLVE: The plot is provided below. By inspection, one might guess that the "lines" that fit these data sets will meet around 1998. This is how the researchers made this leap. CONCLUDE: Men's and women's times have, indeed, grown closer over time. Both sexes have improved their record marathon times over the years, but women's times have improved at a faster rate. In fact, as of 2011, the world record time for men has continued to be faster than the world record time for women. The difference is currently about 686 seconds (under 12 minutes), where in the data plotted, the difference was about 856 seconds. One cannot reasonably extrapolate to years beyond 2010. However, over the past decades, womens' and mens' running times have converged.

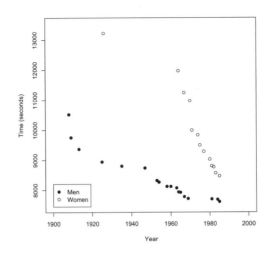

4.45: STATE: Is there a relationship between social distress and brain activity? PLAN: We wish to explore the relationship between social distress and brain activity. We begin with a scatterplot, and compute the correlation if appropriate. SOLVE: A scatterplot shows a fairly strong, positive, linear association. There are no particular outliers; each variable has low and high values, but those points do not deviate from the pattern of the rest. The relationship seems to be reasonably linear. Using software, we compute $r = 0.8782$. CONCLUDE: Social exclusion does appear to trigger a pain response: higher social distress measurements are associated with increased activity in the pain-sensing area of the brain. However, no cause-and-effect conclusion is possible since this was not a designed experiment.

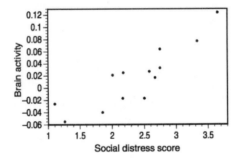

# Chapter 5: Regression

5.1 In this problem, the response variable is "highway mpg," since we're tyring to predict this. The explanatory variable is "city mpg." We're given the regression line:
highway mpg = 6.554 + 1.016 × city mpg. (a) In the equation of the regression line, the slope is the coefficient attached to the explanatory variable ("city mpg" in this case). Hence, here the slope is 1.016. On average, highway mileage increases by 1.016 mile per gallon for each additional 1 mile per gallon increase in city mileage. (b) In the equation of the regression line, the slope is the coefficient not attached to the explanatory variable. Hence, here the intercept is 6.554 mpg. This is the highway mileage for a nonexistent car that gets 0 mpg in the city. Although this interpretation is valid, such a prediction would be invalid, since it involves considerable extrapolation. (c) For a car that gets 16 mpg in the city, we predict highway mileage to be

$$6.554 + (1.016)(16) = 22.81 \text{ mpg}.$$

For a car that gets 28 mpg in the city, we predict highway mileage to be

$$6.554 + (1.016)(28) = 35.002 \text{ mpg}.$$

(d) The regression line passes through all the points of prediction. The plot was created by drawing a line through the two points (16, 22.81) and (28, 35.002), corresponding to the city mileages and predicted highway mileages for the two cars described in (c). We draw the regression line as the line containing these points.

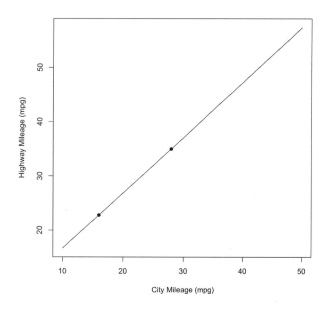

5.3: (a) $\bar{x} = 30.280$, $s_x = 0.4296$, $\bar{y} = 2.4557$, $s_y = 0.1579$, and $r = -0.8914$ Hence,

$$b = r\frac{s_y}{s_x} = (-0.8914)\frac{0.1579}{0.4296} = -0.3276, \text{ and } a = \bar{y} - b\bar{x} = 2.4557 - (-0.3276)(30.280) = 12.3754.$$

(b) Software agrees with these values to three decimal places, since we rounded to the 4th decimal place.

5.5: The farther $r$ is from 0 (in either direction), the stronger the linear relationship is between two variables. The relationship between SRD and DMS is very strongly linear, and a regression line should enable relatively more accurate prediction.

5.7: (a) As in Exercise 5.3, the explanatory variable is mean sea surface temperature ($x$), measured in degrees Celsius, and the response variable is mean coral growth ($y$), measured in mm per year. In Exercise 5.3 above, we computed the least-squares regression line, and found it to be

$$\hat{y} = 12.3754 - 0.3276x.$$

To compute a residual, compare the observed value $y$ to its predicted value, $\hat{y}$. For example, with the first observation ($x = 29.68$, $y = 2.63$), the predicted value $\hat{y}$ is

$$\hat{y} = 12.3754 - (0.3276)(29.68) = 2.652 \text{ mm}.$$

Hence, the residual for this observation is $y - \hat{y} = 2.63 - 2.652 = -0.022$ mm. We continue, computing each residual in this manner. Results are tabulated below:

| $x$ | $y$ | $\hat{y}$ | $y - \hat{y}$ |
|---|---|---|---|
| 29.68 | 2.63 | 2.652 | −0.022 |
| 29.87 | 2.58 | 2.590 | −0.010 |
| 30.16 | 2.60 | 2.495 | 0.105 |
| 30.22 | 2.48 | 2.475 | 0.005 |
| 30.48 | 2.26 | 2.390 | −0.130 |
| 30.65 | 2.38 | 2.335 | 0.045 |
| 30.90 | 2.26 | 2.253 | 0.007 |
| | | | 0 |

(b) They sum to zero, except for rounding error. (c) From software, the correlation between $x$ and $y - \hat{y}$ is 0.000025, which is zero except for rounding.

5.9: (a) Any point that falls exactly on the regression line will not increase the sum of squared vertical distances (which the regression line minimizes). Thus the regression line does not change. Possible output is shown, below left. Any other line (even if it passes through this new point) will necessarily have a higher total sum of squared prediction errors. The correlation changes (increases) because the new point reduces the relative scatter about the regression line. (b) Influential points are those whose $x$ coordinates are outliers. An example is provided, below right.

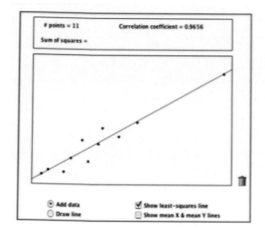 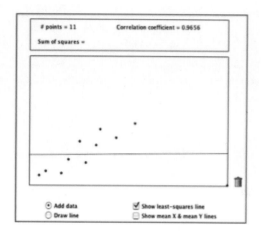

5.11: (a) In the plot, the outlier (Hawaiian Airlines) is the point identified with "H". Since this point is an outlier and falls outside the linear trend suggested by the other data points, it is influential, and will affect the regression line by "pulling" it. (b) With the outlier, $r = -0.624$. If the outlier is deleted from the data, $r = -0.441$. Notice that with the outlier, the correlation suggests a stronger linear relationship. (c) The two regression lines (one including the outlier, and the other without) are plotted. We see that the line based on the full data set (including the outlier) has been pulled down toward the outlier, indicating that the outlier is influential. Now, the regression line based on the complete (original) data set, including the outlier, is $\hat{y} = 27.486 - 0.164x$. Using this, when $x = 74.1$, we predict 15.33% delays. The other regression line (fit without the outlier) is $\hat{y} = 23.804 - 0.069x$, so our prediction would be 18.69% delays. The outlier impacts predictions because it impacts the regression line.

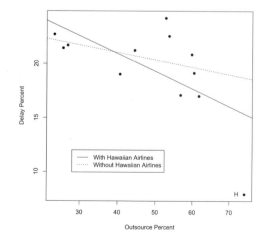

**5.13:** In this problem, we construct the least-squares regression line for predicting the number of manatee deaths (y) from the number of registered boats (x). Software (MINITAB) output is provided, the data are plotted, and the least-squares regression line is added to the plot:

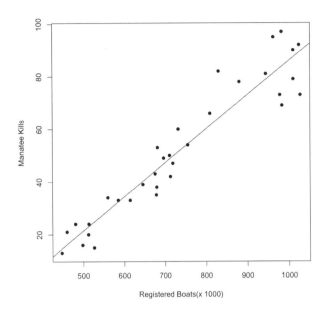

**Minitab output**

The regression equation is ManateeKills = -43.17195 + 0.12923(Boats)

Predictor    Coef      Stdev     t-ratio    p
Constant  -43.17195   5.71582    -7.553    0.0000
Boats       0.12923   0.00752    17.186    0.0000
s = 8.052    R-sq = 90.5%     R-sq(adj) = 90.2%

(a) Reading from the output, the regression line is $\hat{y} = -43.172 + 0.129x$. (b) If 975,000 boats are registered, then by our scale, $x = 975$, and $\hat{y} = -43.172 + (0.129)(975) = 82.6$ manatees killed. The prediction seems reasonable, as long as conditions remain the same, because "975" is within the space of observed values of $x$ on which the regression line was based. That is, this is not extrapolation. (c) If $x = 0$ (corresponding to no registered boats), then we would "predict" –43.172 manatees to be killed by boats. This is absurd, since it is clearly impossible for fewer than 0 manatees to be killed. This illustrates the folly of extrapolation… $x = 0$ is well outside the range of observed values of $x$ on which the regression line was based.

# Solutions

**5.15:** The claim is that children of mothers that smoked during pregnancy scored nine points lower on intelligence tests (on average) at ages three and four than children of nonsmokers. This is obviously an observational study – mothers choose to smoke or not to smoke during pregnancy. With observational studies, cause and effect conclusions are generally not possible because the groups being compared differ in more than one way. In these types of studies there is often a "lurking variable" driving both variables (smoking and intelligence, in this case). Possible lurking variables include the IQ and socioeconomic status of the mother, as well as the mother's other habits (drinking, diet, etc.). These variables are associated with smoking in various ways, and are also predictive of a child's IQ.

**Note:** *There may be an indirect cause-and-effect relationship at work here: some studies have found evidence that over time, smokers lose IQ points, perhaps due to brain damage caused by toxins from the smoke. So, perhaps smoking mothers gradually grow less smart and are less able to nurture their children's cognitive development.*

**5.17:** Age is probably the most important lurking variable: married men would generally be older than single men, so they would have been in the workforce longer, and therefore had more time to advance in their careers.

**5.19:** (b) 0.2. Consider two points on the regression line, say (90,4) and (130,11). The slope of the line segment connecting these points is $\frac{11-4}{130-90} = 7/40$.

**5.21:** (a) $y = 1000 + 100x$

**5.23:** (c) 16 cubic feet.

**5.25:** (a) greater than zero. The slope of the line is positive.

**5.27:** (a) $\hat{y} = 42.9 + 2.5x$

**5.29:** In this problem we predict the price of a diamond ($y$), measured in Singapore dollars, from its size, $x$, measured in Carats. We're given the least-squares regression line: $\hat{y} = 259.63 + 3721.02x$. (a) Since the slope is 3721.02, the least-squares regression line says that increasing the size of a diamond by 1 carat increases its price by 3721.02 Singapore dollars. (b) A diamond of size 0 carats would have a predicted price of 259.63 Singapore dollars. This is probably an extrapolation, since the data set on which the line was constructed almost certainly had no rings with diamonds of size 0 carats. However, if the number is meaningful (dubious), then it refers to the cost of the gold content and other materials in the ring.

**5.31:** In this problem, we predict percent heat loss from the beak (y) from temperature (x). Regression output from MINITAB is provided.

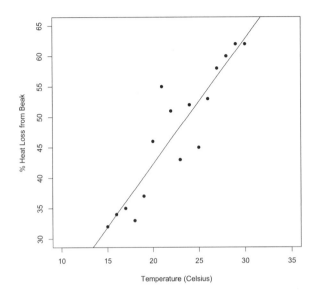

**Minitab output**

The regression equation is PercHeatLoss = 0.919 + 2.065(Temperature)

```
Predictor   Coef      Stdev    t-ratio     p
Constant    0.9191    5.6133   0.164    0.8720
Boats       2.0647    0.2444   8.448    0.0000
s = 4.507   R-sq = 83.6%    R-sq(adj) = 82.4%
```

(a) From the output, the regression equation is $\hat{y} = 0.919 + 2.0647x$. At 25 degrees Celsius, we predict beak heat loss of $\hat{y} = 0.919 + (2.0647)(25) = 52.34$ percent. (b) From the output, $r^2 = 83.6\%$ of the total variation in beak heat loss is explained by the straight-line relationship with temperature. (c) $r = \sqrt{r^2} = \sqrt{0.836} = 0.914$. Correlation is positive here, since the least-squares regression line has a positive slope.

5.33: The $x$-values will be pre-exam scores, and the $y$-values will be final exam scores. This is probably *not* the Princeton University, Nobel prize-winning economist Paul Krugman. (a) $b = r \, s_y/s_x = 0.5 \frac{8}{40} = 0.1$, and $a = \bar{y} - b\bar{x} = 75 - (0.1)(280) = 47$. Hence, the regression equation is $\hat{y} = 47 + 0.1x$. (b) Julie's pre-final exam total was 300, so we would predict a final exam score of $\hat{y} = 47 + (0.1)(300) = 77$. (c) Julie is right… with a correlation of $r = 0.5$, $r^2 = (0.5)^2 = 0.25$, so the regression line accounts for only 25% of the variability in student final exam scores. That is, the regression line doesn't predict final exam scores very well. Julie's score could, indeed, be much higher or lower than the predicted 77. Since she is making this argument, one might guess that her score was, in fact, higher. Julie should visit the Dean.

5.35: (a) The regression equation is $\hat{y} = 28.037 + 0.521x$. $r = 0.555$. (b) The plot is provided. Based on Damien's height of 70 inches, we predict his sister Tonya to have height $\hat{y} = 28.037 + (0.521)(70) = 64.5$ inches (rounded). This prediction isn't expected to be very accurate because the correlation isn't very large… so $r^2 = (0.555)^2 = 0.308$. The regression line explains only 30.8% of the variation in sister heights.

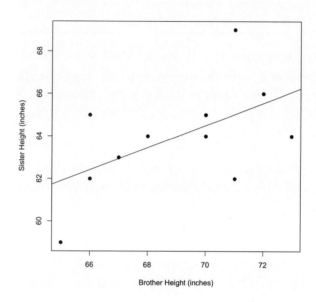

5.37: In this problem, we build a regression line to predict the number of new adults (y) from the percent adult beirds in a colony that return from the previous year (x). The MINITAB output corresponding to this analysis is provided:

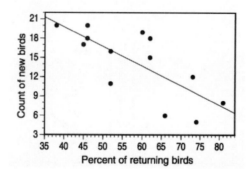

## Minitab output

The regression equation is New = 31.9 − 0.304PctRtn

```
Predictor    Coef      Stdev       t-ratio      p
Constant     31.934    4.838       6.60         0.000
PctRtn       -0.30402  0.0812      -3.74        0.003
s = 3.667    R-sq = 56.0%          R-sq(adj) = 52.0%
```

(a) From the output, the regression equation is $\hat{y} = 31.9 - 0.304x$. The scatterplot and associated regression line are provided above, with the output. (b) The slope (−0.304) tells us that, on the average, for each additional 1% increase in returning birds, the number of new birds joining the colony decreases by 0.304. (c) When $x = 60$ percent, we predict $\hat{y} = 31.9 - 0.304(60) = 13.66$ new birds will join the colony.

5.39: (a) To three decimal places, the correlations are all approximately 0.816 (for Set D, $r$ actually rounds to 0.817), and the regression lines are all approximately $\hat{y} = 3.000 + 0.500x$. For all four sets, we predict $\hat{y} = 8$ when $x = 10$. (b) Plots below. (c) For Set A, the use of the regression line seems to be reasonable—the data seem to have a moderate linear association (albeit with a fair amount of scatter). For Set B, there is an obvious *non*linear relationship; we should fit a parabola or other curve. For Set C, the point (13, 12.74) deviates from the (highly linear) pattern of the other points; if we can exclude it, the (new) regression formula would be very useful for prediction. For Set D, the data point with $x = 19$ is a very influential point—the other points alone give no indication of slope for the line. Seeing how widely scattered the $y$-coordinates of the other points are, we cannot place too much faith in the $y$-coordinate of the influential point; thus we cannot depend on the slope of the line, and so we cannot depend on the estimate when $x = 10$. (We also have no evidence as to whether or not a line is an appropriate model for this relationship.)

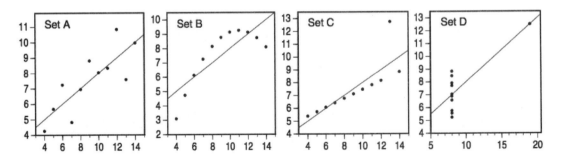

5.41: (a) The regression equation is $\hat{y} = 42.933 + 2.453x$. (b) The regression equation is $\hat{y} = 0.42933 + 0.002453x$. (c) Use the fact that 50 cm = 500 mm. When $x = 50$ cm, the first regression equation gives $\hat{y} = 165.583$ cm. Using the second equation, with $x = 500$ mm, $\hat{y} = 1.65583$ m. These are the same.

5.43: The correlation would be much lower, because there is much greater variation in individuals than in the averages. The correlation in Exercise 4.25 was an ecological correlation, which obscures the variability in individuals.

## Solutions

**5.45:** Responses will vary. For example, students who choose the online course might have more self-motivation or have better computer skills (which might be helpful in doing well in the class; e.g., such students might do better at researching course topics on the Internet).

**5.47:** Here is a (relatively) simple example to show how this can happen: suppose that most workers are currently 30 to 50 years old; of course, some are older or younger than that, but this age group dominates. Suppose further that each worker's current salary is his/her age (in thousands of dollars); for example, a 30-year-old worker is currently making $30,000. Over the next 10 years, all workers age, and their salaries increase. Suppose every worker's salary increases by between $4000 and $8000. Then every worker will be making *more* money than he/she did 10 years before, but *less* money than a worker of that same age 10 years before. During that time, a few workers will retire, and others will enter the workforce, but that large cluster that had been between the ages of 30 and 50 (now between 40 and 60) will bring up the overall median salary despite the changes in older and younger workers.

**5.49:** In this problem we're provided with a scatterplot of first and second-round scores for competitors in the 2010 Masters Tournament. The least-sqaures regression line for predicting the second-round score from a first-round score is given as $\hat{y} = 52.74 + 0.297x$.

For a player who shot 80 in the first round, we predict a second-round score of

$\hat{y} = 52.74 + (0.297)(80) = 76.5$. For a player who shot 70 in the first round, we predict a second-round score of $\hat{y} = 52.74 + (0.297)(70) = 73.53$. Notice that the player who shot 80 the first round (worse than average) is predicted to have a worse-than-average score the second round, but better than the first round. Similarly, the player who shot 70 the first round (better than average) is predicted to do better than average in the second round, but not as well (relatively) as in the first round. Both players are predicted to "regress" to the mean.

**5.51:** See Exercise 4.41 for the three sample scatterplots. A regression line is appropriate only for the scatterplot of part (b). For the graph in (c), the point not in the vertical stack is very influential – the stacked points alone give no indication of slope for the line (if indeed a line is an appropriate model). If the stacked points are scattered, we cannot place too much faith in the y-coordinate of the influential point; thus we cannot depend on the slope of the line, and so we cannot depend on predictions made with the regression line. The curved relationship exhibited by the scatterplot in (d) clearly indicates that predictions based on a straight line are not appropriate.

**5.53:** PLAN: We construct a scatterplot (with beaver stumps as the explanatory variable), and if appropriate, find the regression line and correlation. SOLVE: The scatterplot shows a positive linear association. Regression seems to be an appropriate way to summarize the relationship; the regression line is $=-1.286+11.89x$. The straight-line relationship explains $r^2 =83.9\%$ of the variation in beetle larvae. CONCLUDE: The strong positive association supports the idea that beavers benefit beetles.

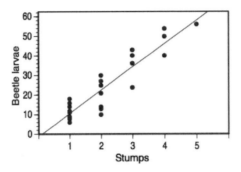

**5.55:** In this problem, we study the relationship between actual number of hurricanes ($y$) and the number forecasted by researchers ($x$). We use a regression line to explore. For perspective, notice that if all data fell on the line $y = x = 0 + 1x$, then all forecasts would be perfectly accurate. Hence, the closer the intercept is to 0 and the slope is to 1, the more accurate the researchers' predictions are. PLAN: We construct a scatterplot, with forecast as the explanatory variable, and Actual as the response variable. If appropriate, we find the least-squares regression line. We consider the impact of the potential outlier (2005 season). SOLVE: A scatterplot is provided below, and the least-squares regression line is included.

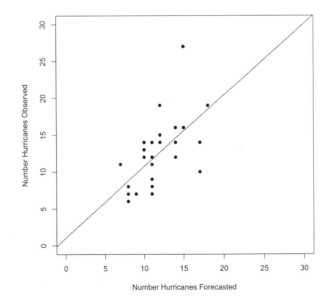

# Solutions

**Minitab output**

The regression equation is Actual = 1.022 + 0.970(Forecasted)

```
Predictor   Coef      Stdev     t-ratio    p
Constant    1.0217    3.0083    0.340      0.7370
PctRtn      0.9696    0.2501    3.876      0.0000
s = 3.794   R-sq = 37.5%    R-sq(adj)=35.0%
```

There is a reasonable, but not very strong linear relationship between Forecasted and Actual hurricanes. In the plot, notice the single observation located farthest from the regression line. Examining the data, this corresponds to the 2005 season, where there were many more hurricanes than researchers forecasted. It is an outlier, and influential, pulling the regression line somewhat. We might consider deleting this point and fitting the line again. Deleting the line, then refitting the model we obtain the regression line, examining the output below, we find $\hat{y} = 2.6725 + 0.7884x$. Deleting the 2005 season, $r = 0.621$, and $r^2 = 38.5\%$. Hence, even after deleting the outlier, the regression line explains only 38.5% of variation in number of hurricanes.

**Minitab output (2005 season deleted)**

The regression equation is Actual = 2.673 + 0.788(Forecasted)

```
Predictor   Coef      Stdev     t-ratio    p
Constant    2.6725    2.4194    1.105      0.2800
PctRtn      0.7884    0.2034    3.877      0.0008
s = 3.006   R-sq = 38.5%    R-sq(adj)=40.0%
```

To demonstrate the impact of removing an outlier, both regression lines are plotted below. Notice how the removal of the outlier (appearing as an open circle) reduces the slope of the regression line.

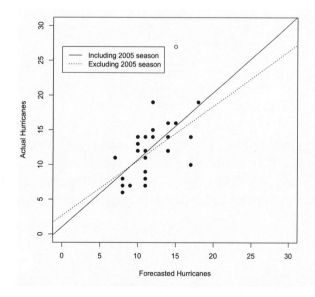

CONCLUDE: Predictions using the regression line are not very accurate. However, there is a positive association, so a forecast of many hurricanes may reasonably be expected to forebode a heavy season for hurricanes.

5.57: We examine how marathon running times are decreasing over time for men and women. We build regression lines for predicting these times ($y$) from year ($x$). PLAN: We plot marathon times by year for each gender, using different symbols. If appropriate, we fit least-squares regression lines for predicting time from year for each gender. We then use these lines to guess when the times will concur. SOLVE: The scatterplot is provided below, with regression lines plotted.

The regression lines are (from software):

For men: $\hat{y} = 67{,}825.3 - 30.44x$

For women: $\hat{y} = 182{,}976.15 - 87.73x$

Although the lines appear to fit the data reasonably well (and the regression line for women would fit better if we omitted the outlier associated with year 1926), this analysis is inviting you to extrapolate, which is never advisable.

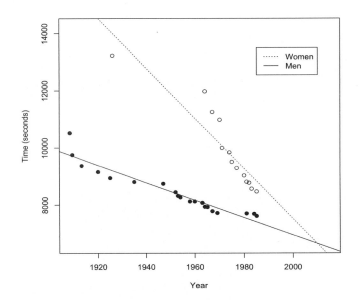

CONCLUDE: Using the regression lines plotted, we might expect women to "outrun" men by the year 2010. Omitting the outlier, the line for women would decrease more steeply, and the intersection would occur sooner, by 1995. Again, such an extrapolation isn't reasonable. Certainly over time, the gap in running times between men and women has decreased markedly.

## Chapter 6: Two-Way Tables

6.1: (a) This table describes $736 + 450 + 193 + 205 + 144 + 80 = 1808$ people. $736 + 450 + 193 = 1379$ played video games. (b) The percent of boys earning A's and B's is $(736+205)/1808 = 0.5205 = 52.05\%$. We do this for all three grade levels. The complete marginal distribution for grades is

| Grade | Percent |
|---|---|
| A's and B's | 52.05% |
| C's | 32.85% |
| D's and F's | 15.10% |

Of all boys, $32.85\% + 15.10\% = 47.95\%$ received a grade of C or lower.

6.3: There are $736 + 450 + 193 = 1379$ players. Of these, $736/1379 = 53.37\%$ earned A's or B's. Similarly, there are $205 + 144 + 80 = 429$ nonplayers. Of these, $205/429 = 47.79\%$ earned A's or B's. Continuing in like manner, the conditional distribution of grades for players follows:

| Grades | Players | Nonplayers |
|---|---|---|
| A's and B's | 53.37% | 47.79% |
| C's | 32.63% | 33.56% |
| D's and F's | 14.00% | 18.65% |

It doesn't look like there's a big difference between these conditional distributions. If anything, players have slightly higher grades (slightly more A's and B's, slightly fewer D's and F's) than nonplayers, but this could be due to chance (more on that later).

6.5: Two examples are shown. In general, choose $a$ to be any number from 10 to 50, and then all the other entries can be determined.

| 30 | 20 |
|---|---|
| 30 | 20 |

| 50 | 0 |
|---|---|
| 10 | 40 |

6.7: (a) For Rotura district, $79/8{,}889 = 0.0089$, or 0.9%, of Maori are in the jury pool, while $258/24{,}009 = 0.0107$, or 1.07%, of the non-Maori are in the jury pool. For Nelson district, the corresponding percents are $1/1{,}328 = 0.08\%$ for Maori and $56/32{,}658 = 0.17\%$ for non-Maori. Hence, in each district, the percent of non-Maori in the jury pool exceeds the percent of Maori in the jury pool. (b) Combining the regions into one table:

|  | Maori | Non-Maori |
|---|---|---|
| In jury pool | 80 | 314 |
| Not in jury pool | 10,138 | 56,353 |
| Total | 10,218 | 56,667 |

For the Maori, overall the percent in the jury pool is $80/10{,}218 = 0.0078$, or 0.78%, while for the non-Maori, the overall percent in the jury pool is $314/56{,}667 = 0.0055$, or 0.55%. Hence, overall the Maori have a larger percent in the jury pool, but in each region they have a lower percent in the jury pool. (c) The reason for Simpson's paradox occurring with this example is that the Maori constitute a large proportion of Rotura's population, while in Nelson they are small minority community.

6.9: (b) 150 teens in schools that forbid cell phones.

6.11: (a) the marginal distribution of school permissiveness.

6.13: (c) the conditional distribution of the frequency that a teen brings a cell phone to school among the schools that forbid cell phones.

6.15: (b) the conditional distribution of school permissiveness among those who brought their cell phone to school every day.

6.17: (b) an example of Simpson's paradox.

6.19 (Expanded): For each type of injury (accidental, not accidental), the distribution of ages is produced below. For example, there were a total of 1,552 accidental weight lifting injuries, of which 239 were experienced by people 19–22 years old. Hence, 239/1,552 = 0.15399, or 15.4%, of accidental injuries were in the 19–22 age group. Similarly, of the 2,559 non-accidental injuries, 916/2,559 = 35.8% were in the 14–18 year age group. Continuing, we obtain the conditional distributions of age for each type of injury:

|       | Accidental | Not accidental |
|-------|------------|----------------|
| 8–13  | 19.0%      | 4.0%           |
| 14–18 | 42.2%      | 35.8%          |
| 19–22 | 15.4%      | 20.8%          |
| 23–30 | 23.4%      | 39.4%          |

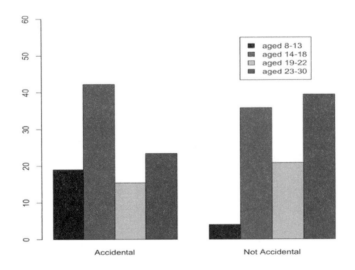

We see that among accidental weight lifting injuries, the percentage of relatively younger lifters is larger, while among the injuries that are not accidental, the percentage of relatively older lifters is larger.

**6.21:** The percent of single men in Grade 1 jobs is 58/337 = 0.172, or 17.2%. The percent of Grade 1 jobs held by single men is 58/955 = 0.0607, or 6.07%.

**6.23:** (a) We need to compute percents to account for the fact that the study included many more married men than single men, so we would expect their numbers to be higher in every job grade (even if marital status had no relationship with job level). (b) A table of percents is provided.

|         | Single | Married | Divorced | Widowed |
|---------|--------|---------|----------|---------|
| Grade 1 | 17.2%  | 11.3%   | 11.9%    | 19.0%   |
| Grade 4 | 2.1%   | 6.9%    | 5.6%     | 9.5%    |

Single and widowed men had higher percents of Grade 1 jobs; single men had the lowest (and widowed men the highest) percents of Grade 4 jobs.

**6.25:** (a) The two-way table of race (White, Black) versus death penalty (Death penalty, no death penalty) follows. Note, for example, that there are 132 + 9 = 141 White defendants that did not receive the death penalty, and 11 + 6 = 17 Black defendants that did receive the death penalty.

|                  | White defendant | Black defendant |
|------------------|-----------------|-----------------|
| Death penalty    | 19              | 17              |
| No death penalty | 141             | 149             |

(b) For black victims: The percentage of white defendants given the death penalty is 0/9 = 0, or 0%. The percentage of black defendants given the death penalty is 6/103 = 0.058, or 5.8%. For white victims: The percentage of white defendants given the death penalty is 19/151 = 0.126, or 12.6%. The percentage of black defendants given the death penalty is 11/63 = 0.175, or 17.5%. Hence, for both victim races, black defendants are given the death penalty relatively more often than white defendants. However, overall, referring to the table in (a), 19/160 = 0.119, or 11.9%, of white defendants got the death penalty, while 17/166 = 0.102, or 10.2%, of black defendants got the death penalty. This illustrates Simpson's paradox. (c) For white defendants, (19 + 132)/(19 + 132 + 0 + 9) = 0.9438 = 94.4% of victims were white. For black defendants, only (11 + 52)/(11 + 52 + 6 + 97) = 0.3795, or 37.95%, of victims were white. Meanwhile, the death penalty was predominantly assigned to cases involving white victims: 14.0% of all cases with a white victim, while only 5.5% of all cases with a black victim had a death penalty assigned to the defendant. Hence, because most white defendants' victims are white, and cases with white victims carry additional risk of a death penalty, white defendants are being assigned the death penalty more often overall.

**6.27: PLAN:** From the given two-way table of results, find and compare the conditional distributions of outcome (success, no success) for each treatment (Chantix, Bupropion and Placebo). **SOLVE:** The percentages for each column are provided in the table. For example, for Chantix, the percentage of successes (no smoking in weeks 9–12) is 155/(155 + 197) = 0.4403, or 44.0%. Since we're comparing success rates, we'll leave off the row for "% smoking in weeks 9–12" since this is just 100%– % no Smoking in weeks 9–12.

|  | Chantix | Bupropion | Placebo |
|---|---|---|---|
| % No smoking in weeks 9–12 | 44.0% | 29.5% | 17.7% |

CONCLUDE: Clearly, a larger percentage of subjects using Chantix were not smoking during weeks 9–12, compared with results for either of the other treatments. In fact, as we'll learn later, this result is statistically significant… random chance doesn't easily explain this difference, and we might conclude that Chantix use increases the chance of success.

6.29: STATE: Is there a difference in the distributions of men and women earning various kinds of degrees? PLAN: We will calculate and compare the conditional distributions of sex for each degree level. SOLVE: For example, the percentage of women earning associate's degrees: 519/823 = 0.631, or 63.1%. The table shows the percent of women at each degree level, which is all we need for comparison. CONCLUDE: Women constitute a substantial majority of associate's, bachelor's, and master's degrees, a slight majority doctor's degrees, and slightly less than 50% of professional degrees.

| Degree | % female |
|---|---|
| Associate's | 63.1% |
| Bachelor's | 57.5% |
| Master's | 61.1% |
| Professional | 49.5% |
| Doctor's | 53.3% |

6.31: STATE: Do smokers and non-smokers differ in their outlooks of their own health? PLAN: We will determine and compare the conditional distributions for health (self-reported) for each group (smokers and non-smokers). That is, we compute the conditional distribution of health outlook for the smokers, then do the same for the non-smokers. SOLVE: The table below provides the percent of subjects with various health outlooks for each group. For example, note that there are 3,906 non-smokers in the survey, of which 1,557 self-reported being in "very good" health. Hence, we have 1,557/3,906 = 0.3986, or 39.9%, of non-smokers self-reporting being in very good health, as depicted in the table below. CONCLUDE: Clearly, the outlooks of current smokers are generally bleaker than that of current non-smokers. Much larger percentages of non-smokers reported being in "excellent" or "very good" health, while much larger percentages of smokers reported being in "fair" or "poor" health.

|  | Health Outlook | | | | |
|---|---|---|---|---|---|
|  | Excellent | Very good | Good | Fair | Poor |
| Current smoker | 6.2% | 28.5% | 35.9% | 22.3% | 7.2% |
| Current non-smoker | 12.4% | 39.9% | 33.5% | 14.0% | 0.3% |

# Chapter 7: Exploring Data: Part I Review

Test Yourself Exercise Answers are sketches. All of these problems are similar to ones found in Chapters 1–6, for which the solutions in this manual provide more detail.

7.1: (c)

7.3: (c)

7.5: (b)

7.7: (a)

7.9: (d)

7.11: (b)

7.13: (b)

7.15: (a)

7.17: $P(X > 90) = P\left(Z > \dfrac{90 - 75}{8.3}\right) = P(Z > 1.81)$. From Table A, the area under the standard Normal curve to the left of 1.81 is 0.9649. Hence, $P(X > 90) = 1 - 0.9649 = 0.0351$, or 3.51%. (b) The middle 50% of all observations lie between the first and third quartiles, so the IQR is the range in which these observations lie. To find the first quartile, using Table A, find a $z$ score corresponding to left-tail area of 0.2500 (look up 0.2500 as an area). The first quartile has $z = -0.67$. By symmetry, with area 0.2500 in the right tail, the third quartile has $z = 0.67$. Hence, the first and third quartiles are, respectively, $Q_1 = 75 - 0.67(8.3) = 69.44$ and $Q_3 = 75 + 0.67(8.3) = 80.56$ ksi. The range (IQR) in which the middle values lie is therefore $80.56 - 69.44 = 11.12$ ksi.

7.19: (a) Sort the data first:

7.2  7.6  8.5  8.5  8.7  9.0  9.0  9.3  9.4  9.4  10.2  10.9  11.3  12.1  12.8

We see that the 5-number summary is given by Minimum = 7.2, $Q_1$ = 8.5, Median = 9.3, $Q_3$ = 10.9, Maximum = 12.8. (b) Examine the boxplot provided in Figure 7.3. The median corresponds to the "middle" line in the box of the boxplot. Here, Median is closest to 27. (c) The third quartile corresponds to the upper-end of the box in the boxplot. We have that 25% of values exceed $Q_3 = 30$. (d) Yes. Virtually all Torrey pine needles are longer than virtually all Aleppo pine needles. There is no overlap in the distributions, as seen by comparing, say, Minimum for Torrey pine needles (21) to Maximum for Aleppo pine needles (12.8).

7.21: (b)

7.23: (c)

7.25: (c) The regression line's intercept is the predicted value of $y$ when $x = 0$. We would predict 1.41 offspring per female when the cone index ($x$) is 0.

7.27: (d) If people at age 55 could remember the diets they had when they were 18 years old, we would expect a very strong and positive linear relationship between the values they report at age 18 and the values they report when they are 55. Here the reported correlation is actually negative. If anything, this suggests that there is virtually no relationship between calories reported at age 18 and calories reported at age 55.

7.29: (d)

7.31: (c)

7.32: (b)

7.33: (a)

7.35: (a) No. (b) $r^2 = 0.64$, or 64%.

7.37: (a) For lean monkeys, the mean lean body mass is 8.683 kg. (b) For obese monkeys, the mean lean body mass is 10.517 kg. (c) Such a comparison would be unreasonable because the lean group is less massive, and therefore would be expected to burn less energy on average. (d) A scatterplot follows, where lean and obese monkeys are indicated with "L" and "O", respectively. (e) Based on the plot, it appears that the rate of increase in energy burned per kilogram of mass is about the same for both groups. Of course, the obese monkeys are more massive, and therefore, on average, burn more energy, as computed in (a) and (b).

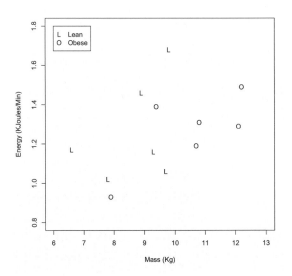

7.39: (a) $190/8474 = 0.0224$, or 2.24%. (b) $633/8474 = 0.0747$, or 7.47%. (c) $27/633 = 0.0427$, or 4.27%. (d) $4621/8284 = 0.5578$, or 55.78%. (e) The conditional distribution of CHD for each level of anger is tabulated below. The result for the high anger group was computed in (c), for example. Clearly, angrier people are at greater risk of CHD.

| Low anger | Moderate anger | High anger |
|---|---|---|
| 1.70% | 2.33% | 4.27% |

7.41: The time plot shows a lot of fluctuation from year to year, but also shows a recent increase: Prior to 1972, the discharge rarely rose above 600 km$^3$, but since then, it has exceeded that level more than half the time. A histogram or stemplot cannot show this change over time because those plots use information on only one variable (discharge rate), but not on year.

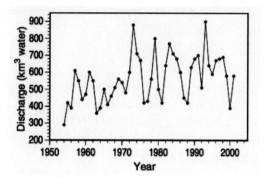

7.43: (a) The plot is provided. (b) The least-squares regression line is $\hat{y} = 160.79 - 0.07410x$. The slope is negative, suggesting that the ice breakup day is decreasing (by 0.07410 day per year). (c) The regression line is not very useful for prediction, as it accounts for only about 11.7% ($r^2 = 0.117$) of the variation in ice breakup time.

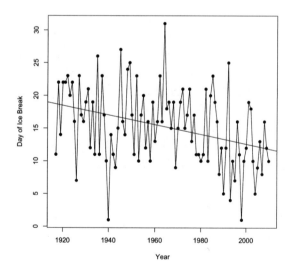

7.45: (a) and (b) Two stemplots are provided. The first shows all the data points, and the second omits the three highest countries, which are identified as outliers by use of the 1.5 × IQR criterion. In the absence of those outliers, the distribution is roughly symmetric. (c) The mean and standard deviation are $\bar{x} = 18.2447\%$ and $s = 7.0451\%$. Some students may instead report the five-number summary, which is Min = 3.14%, $Q_1 = 13.64\%$, $M = 18.27\%$, $Q_3 = 24.08\%$, Max = 34.83%, but this would be appropriate for a very skewed distribution (such as the case without omitting outliers). (d) The U.S. share of G.D.P. is small compared to the other countries in this list; more than one standard deviation below the mean and below the first quartile of the distribution.

```
 0 | 3456889
 1 | 000011112223333444444445556666778888889999999
 2 | 000111334444445666899999
 3 | 044
 4 | 8
 5 |
 6 | 3
 7 |
 8 |
 9 |
10 | 6
```

```
0 | 3
0 | 45
0 | 6
0 | 889
1 | 00001111
1 | 2223333
1 | 44444444555
1 | 666677
1 | 888888999999
2 | 000111
2 | 33
2 | 44444445
2 | 666
2 | 899999
3 | 0
3 |
3 | 44
```

7.47: STATE: How does angle of deformity vary among young HAV patients requiring surgery? PLAN: Display the distribution with a graph and compute appropriate numerical summaries. SOLVE: A stemplot is shown; a histogram could also be used. The distribution seems to be fairly Normal apart from a high outlier of 50°. The five-number summary is preferred because of the outlier: Min = 13°, $Q_1 = 20°$, $M = 25°$, $Q_3 = 30°$, Max = 50°. Alternatively, the mean and standard deviation are $\bar{x} = 25.4211°$ and $s = 7.4748°$. CONCLUDE: Student descriptions of the distribution will vary. Most patients have a deformity angle in the range of 15° to 35°.

```
1 | 34
1 | 66788
2 | 000111123
2 | 55556666888
3 | 00012224
3 | 88
4 |
4 |
5 | 0
```

7.49: STATE: Can severity of MA be used to predict severity of HAV? PLAN: We examine the relationship with a scatterplot and (if appropriate) correlation and regression line. SOLVE: MA angle is the explanatory variable, so it should be on the horizontal axis of the scatterplot. The scatterplot shows a moderate to weak positive linear association, with one clear outlier (the patient with HAV angle 50°). The correlation is $r = 0.3021$, and the regression line is $\hat{y} = 19.723 + 0.3388x$. CONCLUDE: MA angle can be used to give (very rough, imprecise) estimates of HAV angle, but the spread is so wide that the estimates would not be very reliable. The linear relationship explains only $r^2 = 9.1\%$ of the variation in HAV angle.

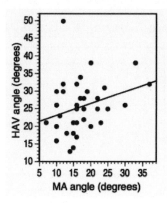

7.51: STATE: How does the cylinder wall thickness influence the gate velocity chosen by the skilled workers? PLAN: We will examine the relationship with a scatterplot and (if appropriate) correlation and regression line. We then interpret this output to address the question. SOLVE: The scatterplot of Gate velocity versus Cylinder wall thickness, shown with the regression line $\hat{y} = 70.44 + 274.78x$, shows a moderate, positive linear relationship. The linear relationship explains about $r^2 = 49.3\%$ of the variation in gate velocity. CONCLUDE: The regression formula might be used as a rule of thumb for new workers to follow, but the wide spread in the scatterplot suggests that there may be other factors that should be taken into account in choosing the gate velocity.

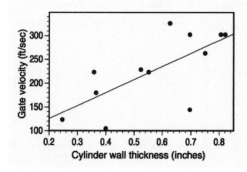

7.53: (a) The scatterplot of 2003 returns against 2002 returns shows (ignoring the outlier) a strong negative association. (b) The correlation for all 23 points is $r = -0.6230$; with the outlier removed, the correlation is $r = -0.8722$. The outlier deviates from the linear pattern of the other points; removing it makes the negative association stronger, and so $r$ moves closer to $-1$.

(c) Regression formulas are given in the table on the right. The first line is solid in the plot, the second is the dashed line. The least-squares regression line makes the sum of the squares of the vertical deviations of the points from the line as small as possible. The line for the 22 other funds is so far below Fidelity Gold that the squared deviation is very large. The line must pivot up toward Fidelity Gold in order to minimize the sum of squares for all 23 deviations. Fidelity Gold is very influential.

|  | $r$ | Equation |
|---|---|---|
| All 23 funds | $-0.6230$ | $\hat{y} = 29.2512 - 0.4501x$ |
| Without Fidelity Gold | $-0.8722$ | $\hat{y} = 18.1106 - 0.9429x$ |

7.55: (a) Fish catch (on the horizontal axis) is the explanatory variable. The point for 1999 is at the bottom of the plot. (b) The correlations are given in the table below. The outlier decreases $r$ because it weakens the strength of the association. (c) The two regression lines are given in the table; the solid line in the plot uses all points, while the dashed line omits the outlier. The effect of the outlier on the line is small: it pulls the line down on the left side (and increases the slope) very slightly, but for making predictions, both lines would give similar results.

|  | $r$ | Equation |
|---|---|---|
| All points | 0.6724 | $\hat{y} = -21.09 + 0.6345x$ |
| Without 1999 | 0.8042 | $\hat{y} = -19.05 + 0.5788x$ |

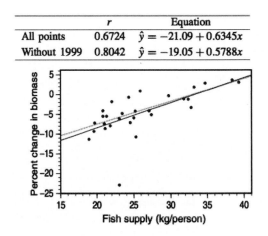

# Chapter 8: Producing Data: Sampling

8.1: (a) The population is (all) college students. (b) The sample is the 104 students at the researcher's college who returned the questionnaire.

8.3: (a) The population is all 45,000 people who made credit card purchases. (b) The sample is the 137 people who returned the survey form.

8.5: Since all the students surveyed are enrolled in a special senior honors psychology class, these students may be more likely to be interested in joining the club (and more willing to pay $35 to do so). The direction of bias is likely to overestimate the proportion of all psychology majors willing to pay to join this club. This is a convenience sample.

8.7: There are 26 managers in the population. Number these from 01 to 26 alphabetically (down the columns): Adelaja = 01, Ahmadiani = 02, ..., Yajima = 26. If you use the applet: Set Population = 1 to 26, select a sample of size 5, then click Reset and Sample. If you use Table B, enter at line 134 and choose 16 = Ippolito, 18 = Jung, 13 = Gupta, 21 = Modur, and 04 = Bonds.

8.9: With the election close at hand, the polling organization wants to increase the accuracy of its results. Larger samples provide better information about the population.

8.11: Label the suburban townships from 01 to 30, down the columns. With Table B, enter at line 105 and choose 29 = Wheeling, 07 = Elk Grove, 19 = Orland, 14 = New Trier, and 17 = Norwood Park. Next, label the Chicago townships from 1 to 8, down the columns. With Table B, enter at line 115 and choose 6 = Rogers Park, 1 = Hyde Park, and 4 = Lake View.

8.13: The higher no-answer was probably the second period—more families are likely to be gone for vacations, or to be outside enjoying the warmer weather, and so on. Nonresponse of this type might underrepresent those who are more affluent (and are able to travel). In general, high nonresponse rates always make results less reliable, because we do not know what information we are missing.

8.15: (a) and (b) Features will vary depending on the website chosen. (c) The weakness of any online poll is that it relies on voluntary response. Most online poll samples are not representative of any larger population of use or interest to the researcher.

8.17: (a) all customers who have purchased something in the last year.

8.19: (b) 5458, 0815, 0727, 1025, 6027.

8.21: (b) a stratified random sample (plots are stratified by terrain).

8.23: (c) 04, 18, 07, 13, 02, 05. (Notice that in (b) "07" appears in the sample twice.)

8.25: (b) The result for the entire sample is more accurate because both come from a larger sample. The sample of Republicans is a subset of the entire sample – there are fewer Republicans than people surveyed overall.

8.27: The population is the 1,000 envelopes stuffed during a given hour. The sample is the 40 envelopes selected.

8.29: With the applet: Population = 1 to 287, select a sample of size 20, then click Reset and Sample. Using Table B, number the area codes 001 to 287. Then, enter at line 135, and pay attention to the instructions that if we use the table, we'll pick only 5 numbers. The selected area codes are 255, 100, 120, 126, 008.

8.31: (a) Alphabetize the 6168 names (using middle initials or a student ID to distinguish between two people with the same name). Label these students with an ID 0001 to 6168. (b) Using Table B, entering at line 135, the sample is 5556, 5839, 1007, 1120, 1513, 1260, 0842, and 1447.

8.33: (a) False. Such regularity holds only in the long run. If it were true, you could look at the first 39 digits and know whether or not the 40th digit was a 0. (b) True. All pairs of digits (there are 100, from 00 to 99) are equally likely. (c) False. Four random digits have chance 1/10,000 to be 0000, so this sequence will occasionally occur. The sequence 0000 is no more or less random than 1234 or 2718, or any other four-digit sequence.

8.35: Online polls, call-in polls, and voluntary response polls in general tend to attract responses from those who have strong opinions on the subject, and therefore are often not representative of the population as a whole. On the other hand, there is no reason to believe that randomly chosen adults would over-represent any particular group, so the responses from such a group give a more reliable picture of public opinion.

8.37: (a) Assign labels 0001 through 5024, enter the table at line 104, and select: 1388, 0746, 0227, 4001, and 1858. (b) More than 171 respondents have run red lights. We would not expect very many people to claim they *have* run red lights when they have not, but some people will deny running red lights when they have.

8.39: (a) Each person has a 10% chance of being selected: 4 of 40 men, and 3 of 30 women. (b) This is not an SRS because not every group of 7 people can be chosen; the only possible samples are those with 4 men and 3 women.

8.41 (Exapanded): Sample separately in each stratum; that is, assign separate labels, then choose the first sample, then continue on in the table to choose the next sample, etc. Beginning with line 102 in Table B, we choose:

| Forest type | Labels | Parcels selected |
|---|---|---|
| Climax 1 | 01 to 36 | 19, 27, 26, 17 |
| Climax 2 | 01 to 72 | 09, 55, 32, 22, 69, 56, 52 |
| Climax 3 | 01 to 31 | 13, 07, 02 |
| Secondary | 01 to 42 | 27, 40, 01, 18 |

*Solutions* 51

8.43: (a) Since 200/5 = 40, we will choose one of the first 40 names at random. Beginning on line 120, the addresses selected are 35, 75, 115, 155, and 195. (Only the first number is chosen from the table.) (b) All addresses are equally likely; each has chance 1/40 of being selected. To see this, note that each of the first 40 has chance 1/40 because one is chosen at random. But each address in the second 40 is chosen exactly when the corresponding address in the first 40 is, so each of the second 40 also has chance 1/40. And so on. This is not an SRS because the only possible samples have exactly one address from the first 40, one address from the second 40, and so on. An SRS could contain any 5 of the 200 addresses in the population. Note that this view of systematic sampling assumes that the number in the population is a multiple of the sample size.

8.45: (a) Automated random digit dialing is a fast, economical way to randomly dial landline telephone numbers. (b) In some families, the adult that answers the phone regularly may be systematically different from an adult that does not. For example, people not working at a job may be at home more often, and therefore may be more likely to answer the phone. (c) There could be (and probably are) big differences between landline phone users and cellular phone users. The design in question is a stratified sample.

8.47: Answers will vary considerably. See the textbook for several examples. (a) One example: "Should colleges do away with the 'tenure' system, which effectively allows lazy and incompetent faculty members to stay in highly-paid, easy, taxpayer-funded jobs?" (b) On example: "Do you regularly look at online pornography?"

# Chapter 9: Producing Data: Experiments

9.1: This is an observational study: No treatment was assigned to the subjects; we merely observed cell phone usage (and presence/absence of cancer). The explanatory variable is cell phone usage, and the response variable is whether or not a subject has brain cancer.

9.3: This is an observational study, so it is not reasonable to conclude any cause-and-effect relationship. At best, we might advise smokers that they should be mindful of potential weight gain and its accompanying ailments.

9.5: Individuals: pine seedlings. Factor: amount of light. Treatments: full light, 25% light, or 5% light. Response variable: dry weight at the end of the study.

9.7: Making a comparison between the treatment group and the percent finding work *last year* is not helpful. Over a year, many things can change: the state of the economy, hiring costs (due to an increasing minimum wage or the cost of employee benefits), etc. (In order to draw conclusions, we would need to make the $500 bonus offer to some people and not to others, and compare the two groups.)

9.9: (a) We assign 18 plots randomly to the three groups; 6 plots to each group. Group 1 will have water added in January to March; Group 2 has water added in April to June; Group 3 is the control group and has no water added. Measurements are taken, and the three groups are compared. See the Diagram below. (b) If using Table B, label 01 to 36 and take two digits at a time, as seen in Chapter 8. For example, suppose the first numbers in Table B are 01329 31746 11084 31262 88401 07233. Then we assign to Group 1 plots 01, 17, 11, 08, 12 and 10.

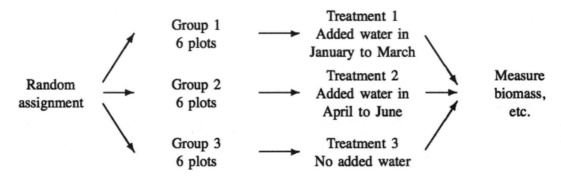

9.11: In a controlled scientific study, the effects of factors other than the nonphysical treatment (e.g., the placebo effect, differences in the prior health of the subjects) can be eliminated or accounted for, so that the differences in improvement observed between the subjects can be attributed to the differences in treatments.

9.13: (a) The researchers simply observed the diets of subjects; they did not alter them. (That is, no treatments were assigned.) (b) Such language is reasonable because with observational studies, no "cause and effect" conclusion would be reasonable.

*Solutions*

9.15: In this case, "lack of blindness" means that the experimenter knows which subjects were taught to meditate. He or she may have some expectation about whether or not meditation will lower anxiety; this could unconsciously influence the diagnosis.

9.17: (a) *Completely randomized design:* Randomly assign 15 students to Group 1 (easy mazes) and the other 15 to Group 2 (hard mazes). Compare the time estimates of Group 1 with those of Group 2. (b) *Matched-pairs design:* Each student does the activity twice, once with the easy mazes, and once with the hard mazes. Randomly decide (for each student) which set of mazes is used first. Compare each student's "easy" and "hard" time estimate (for example, by looking at each "hard" minus "easy" difference). *Alternate matched-pairs design:* Again, all students do the activity twice. Randomly assign 15 students to Group 1 (easy first) and 15 to Group 2 (hard first).

9.19: (a) This is an observational study: behavior (alcohol consumption) is observed, but no treatment is imposed.

9.21: (c) two factors, each with two levels.

9.23: (b) the score on the memory test of their recall of advertisements is the response.

9.25: (b) The communities are paired up, then one is chosen to have the advertising campaign.

9.27: (b) This was a (matched-pairs) experiment, but in order to give useful information, the subjects should be chosen from those who might be expected to buy this car.

9.29: This is an experiment, because the treatment is selected (randomly, we assume) by the interviewer. The explanatory variable (treatment) is the level of identification, and the response variable is whether or not the interview is completed.

9.31: (a) In an observational study, we simply observe subjects who have chosen to take supplements and compare them with others who do not take supplements. In an experiment, we *assign* some subjects to take supplements and assign the others to take no supplements (or better yet, assign the others to take a placebo). (b) "Randomized" means that the assignment to treatments is made randomly, rather than by some other method (e.g., asking for volunteers). "Controlled" means that some subjects were used as a "control" group—probably meaning that they received placebos—which gives a basis for comparison to observe the effects of the treatment. (c) Subjects who choose to take supplements have other characteristics that are confounded with the effect of the supplements; one of those characteristics is that people in this group are more likely to make healthy lifestyle choices (about smoking, drinking, eating, exercise, etc.). When we randomly assign subjects to a treatment, the effect of those characteristics is erased, because some of those subjects will take the supplement, and some will take the placebo.

9.33: (a) We assign the 120 schools randomly to two treatment groups. Treatment group 1 uses the camera, and Treatment group 2 serves as a control, with no camera. Attendance is compared between the two groups. See the diagram. (b) Assign labels 001 to 120. If using Table B, line 108 gives 090, 009, 067, 092, 041, 059, 040, 080, 029, 091 as part of the sample assigned to Treatment group 1. When 60 schools have been selected, the remaining schools are allocated to Treatment group 2.

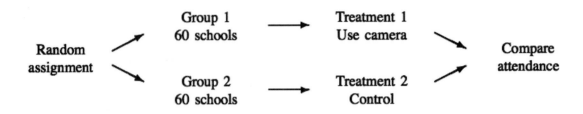

9.35: Use a completely randomized design; the diagram is provided. Labeling the men from 01 through 39, and starting on line 107 of Table B, we make the assignments shown in the table on the right.

Group 1:  20, 11, 38, 31, 07, 24, 17, 09, 06
Group 2:  36, 15, 23, 34, 16, 19, 18, 33, 39
Group 3:  08, 30, 27, 12, 04, 35
Group 4:  02, 32, 25, 14, 29, 03, 22, 26, 10
Group 5:  Everyone else

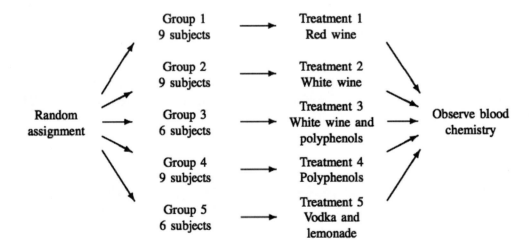

9.37: (a) The outline is given below. There are 40 subjects, so we assign 10 subjects to each of the four treatments. The four treatments are outlined:

|  | Antidepressant | No drug |
|---|---|---|
| Stress management | 1 | 2 |
| None | 3 | 4 |

(b) Assign labels 01 through 40 (in alphabetical order). The full randomization is easy with the Simple Random Sample applet: each successive sample leaves the population hopper, so that you need only click Sample three times to assign 30 subjects to three groups; the 10 subjects remaining in the hopper are the fourth group. Alternatively, line 125 of Table B gives the following subjects for Group 1: 21 Jiang, 37 Suarez, 18 Hersch, 23 Kim, 19 Hurwitz, 03 Alawi, 39 Wilson   24 Landers, 27 Morgan, and 13 Garrett.

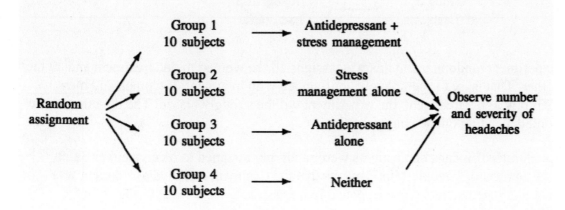

**9.39:** The factors are pill type and spray type. "Double-blind" means that the treatment assigned to a patient was unknown to both the patient and those responsible for assessing the effectiveness of that treatment. "Placebo-controlled" means that some of the subjects were given placebos. Even though these possess no medical properties, some subjects may show improvement or benefits just as a result of participating in the experiment; the placebos allow those doing the study to observe this effect.

**9.41:** (a) The subjects are randomly chosen Starbucks customers. Each subject tastes two cups of coffee, in identical unlabeled cups. One contains regular mocha frappuccino, the other the new light version. The cups are presented in random order, half the subjects get regular then light, the other half light then regular. Each subject says which cup he or she prefers. (b) We must assign 10 customers to get regular coffee first. Label the subjects 01 to 20. Starting at line 141, the "regular first" group is: 12, 16, 02, 08, 17, 10, 05, 09, 19, 06.

**9.43:** Each player will be put through the sequence (100 yards, four times) twice—once with oxygen and once without. For each player, randomly determine whether to use oxygen on the first or second trial. Allow ample time (perhaps a day or two) between trials for full recovery.

**9.45:** Here we assign the 300 subjects randomly to three treatment groups. Group 1 receives just wine; Group 2 receives just beer, and Group 3 receives just spirits. A (blinded) professional measures the heart health of all patients, and the groups are compared. The diagram is shown below.

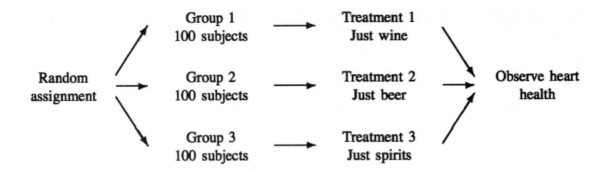

**9.47:** Any experiment randomized in this way assigns all the women to one treatment and all the men to the other. That is, sex is completely confounded with treatment. If women and men respond differently to the treatment, the experiment will be strongly biased. The direction of the bias is random, depending on the coin toss.

**9.49:** (a) "Randomized" means that patients were randomly assigned to receive either Saint-John's wort or a placebo. "Double-blind" means that the treatment assigned to a patient was unknown to both the patient and those responsible for assessing the effectiveness of that treatment. "Placebo-controlled" means that some of the subjects were given placebos. Even though these possess no medical properties, some subjects may show improvement or benefits just as a result of participating in the experiment; the placebos allow those doing the study to observe this effect. (b) Diagram below. There isn't a problem with having two different sample sizes in the two groups being compared, as long as both sample sizes are reasonably large.

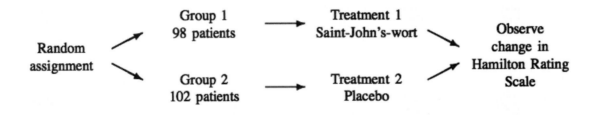

## Chapter 10: Introducing Probability

10.1: In the long run, of a large number of Texas Hold 'em games in which you hold a pair, the fraction in which you can make four of a kind will be about 2/245. It *does not* mean that exactly 2 out of 245 such hands would yield four of a kind. The probability of an event is the long-run frequency of times the event occurs if the experiment is repeated endlessly… not 245 times.

10.3: (a) There are 21 zeros among the first 200 digits of the table (rows 101–105, combined), for a proportion of 0.105. (b) Answers will vary, but more than 99% of all students should get between 7 and 33 heads out of 200 flips.

10.5: (a) $S$ = {lives on campus, lives off campus}. (b) $S$ = {All numbers between $A$ and $B$ years}. (Choices of upper and lower limits $A$ and $B$ will vary.) (c) $S$ = {all amounts greater than or equal to 0}, or $S$ = {0, 0.01, 0.02, 0.03, …}. (d) $S$ = {A, B, C, D, F} (students might also include "+" and "–").

10.7: For the sample space, add 1 to each pair-total possible in rolling a four-sided die twice. For example, if you roll two 1's, then Intelligence takes value $1 + 1 + 1 = 3$. If you roll a 4 and 3, then Intelligence = $4 + 3 + 1 = 8$. We see that $S$ = {3, 4, 5, 6, 7, 8, 9}. As all faces are equally likely and the dice are independent, each of the 16 possible pairings is equally likely, so (for example) the probability of a total of 5 is 3/16, because Intelligence is 5 if we roll (1, 3), (3, 1), or (2,2). There are therefore 3 pairings leading to a value of Intelligence of 5. Similarly, Intelligence of 6 occurs if we roll (2,3), (3,2), (4,1), or (1,4), so the probability of this is 4/16. The complete set of probabilities is shown in the table.

| Total | Probability |
|-------|-------------|
| 3 | 1/16 |
| 4 | 2/16 |
| 5 | 3/16 |
| 6 | 4/16 |
| 7 | 3/16 |
| 8 | 2/16 |
| 9 | 1/16 |

10.9: (a) Event $B$ specifically rules out obese subjects, so there is no overlap with event $A$. (b) $A$ or $B$ is the event "The person chosen is overweight or obese." $P(A \text{ or } B) = P(A) + P(B) = 0.34 + 0.33 = 0.67$. (c) $P(C) = 1 - P(A \text{ or } B) = 1 - 0.67 = 0.33$.

10.11: Model 1: Not legitimate (probabilities have sum 6/7). Model 2: Legitimate. Model 3: Not legitimate (probabilities have sum 7/6). Model 4: Not legitimate (probabilities cannot be more than 1).

10.13: (a) This is a legitimate probability model because the probabilities sum to 1. (b) The event $\{X < 4\}$ is the event that somebody lifts weights 3 or fewer days per week. $P(X<4) = 0.73 + 0.06 + 0.06 + 0.06 = 0.91$. (c) This is the event $\{X \geq 1\}$. $P(X \geq 1) = 1 - P(X = 0) = 1 - 0.73 = 0.27$.

10.15: (a) The area of a triangle is $\frac{1}{2}bh$. Here, $b = 2$ and $h = 1$. So, Area = $\frac{1}{2}(2)(1) = 1$. (b) We need the area under the density curve below 1. This is illustrated in the diagram below, on the left. $P(X < 1) = 0.5$, since this is exactly half the total area under the density curve (half of 1). (c) We need the area under the density curve below 0.5. This is illustrated in the diagram below, on the right. We have $P(X < 0.5) = \frac{1}{2}bh = \frac{1}{2}(0.5)(0.5) = 0.125$.

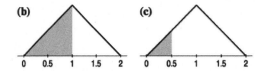

10.17: (a) $X \geq 3$ means the student's grade is B or higher (B, B+, A– or A). $P(X \geq 3) = P(3) + P(3.3) + P(3.7) + P(4.0) = 0.14 + 0.10 + 0.08 + 0.09 = 0.41$. (b) "Poorer than B–" means any grade **lower** than B– (e.g. C+, C, C–, D+, D or F). We want $P(X < 2.7) = P(X \leq 2.3) = P(C+) + P(C) + P(C-) + P(D+) + P(D) + P(F) = 0.10 + 0.12 + 0.04 + 0.04 + 0.08 + 0.08 = 0.46$.

10.19: (a) Answers will vary (probably wildly). (b) A personal probability might take into account specific information about one's own driving habits, or about the kind of traffic one usually drives in. (c) Most people believe that they are better-than-average drivers (whether or not they have any evidence to support that belief).

10.21: (a) Probabilities express the *approximate* fraction of occurrences out of many trials.

10.23: (b) This is a discrete (but not equally likely) model.

10.25: (c) $P$(Republican or Democrat) = $P$(Republican) + $P$(Democrat) = $0.28 + 0.28 = 0.56$.

10.27: (b) There are 10 equally likely possibilities, so $P$(seven) = 1/10.

10.29: (b) 24% (0.16 + 0.05 + 0.02 + 0.01 = 0.24, or 24%) have 3 or more cars.

10.31: (a) There are sixteen possible outcomes: { HHHH, HHHM, HHMH, HMHH, MHHH, HHMM, HMHM, HMMH,MHHM, MHMH, MMHH, HMMM, MHMM, MMHM, MMMH, MMMM } (b) The sample space is {0,1,2,3,4}.

10.33: (a) The given probabilities have sum 0.73, so this probability must be $1 - 0.73 = 0.27$. (b) $P$(at least a high school education) = $1 - P$ (has not finished HS) = $1 - 0.13 = 0.87$. (Or, add the other three probabilities.)

10.35: (a) All probabilities are between 0 and 1, and they add to 1. (We must assume that no one takes more than one language.) (b) The probability that a student is studying a language other than English is $0.43 = 1 - 0.57$ (or add all the other probabilities). (c) This probability is $0.40 = 0.30 + 0.08 + 0.02$.

Solutions

**10.37:** Of the seven cards, there are three 9's, two red 9's, and two 7's. (a) $P$(draw a 9) = 3/7. (b) $P$(draw a red 9) = 2/7. (c) $P$(don't draw a 7) = 1 − $P$(draw a 7) = 1 − 2/7 = 5/7.

**10.39:** Each of the 90 guests has probability 1/90 of winning the prize. The probability that the winner is a woman is the sum of 1/90 42 times, one for each woman. The probability is 42/90 = 0.467.

**10.41:** (a) It is legitimate because every person must fall into exactly one category, the probabilities are all between 0 and 1, and they add up to 1. (b) $P$(15–19-year-old with others) = 0.169. (c) $P$(15–19-year-old) = 0.171—the sum of the numbers in the first column. (d) $P$(lives with others) = 0.532—the sum of the numbers in the third row.

**10.43:** (a) $P$(20 years old or older) = 1 − 0.171 = 0.829 (or sum the entries in the second, third and fourth columns). (b) $P$(does not live alone) = 1 − $P$(lives alone) = 1 − 0.073 = 0.927.

**10.45:** (a) All 9 digits are equally likely, so each has probability 1/9:

| Value of W | 1 | 2 | 3 | 4 | 5 | 6 | 7 | 8 | 9 |
|---|---|---|---|---|---|---|---|---|---|
| Probability | $\frac{1}{9}$ | $\frac{1}{9}$ | $\frac{1}{9}$ | $\frac{1}{9}$ | $\frac{1}{9}$ | $\frac{1}{9}$ | $\frac{1}{9}$ | $\frac{1}{9}$ | $\frac{1}{9}$ |

(b) $P(W \geq 6) = P(W=6) + P(W=7) + P(W=8) + P(W=9) = 4/9 = 0.444$, or twice as big as the Benford's law probability.

**10.47:** (a) BBB, BBG, BGB, GBB, GGB, GBG, BGG, GGG. Each has probability 1/8. (b) Three of the eight arrangements have two (and only two) girls, so $P(X = 2) = 3/8 = 0.375$. (c) See table.

| Value of X | 0 | 1 | 2 | 3 |
|---|---|---|---|---|
| Probability | 1/8 | 3/8 | 3/8 | 1/8 |

**10.49:** (a) This is a continuous random variable because the set of possible values is an interval. (b) The height should be 1/2 because the area under the curve must be 1. The density curve is illustrated. (c) $P(Y \leq 1) = 1/2$.

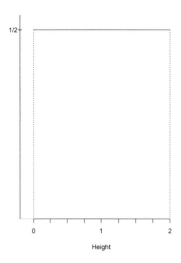

**10.51:** $V$ denotes the probability that a randomly selected voter voted for Barack Obama. Then $V$ has (approximately) the Normal distribution with mean 0.53 and standard deviation 0.009. (a) $P(0.51 \leq V \leq 0.55) = P(\frac{0.51-0.53}{0.009} \leq Z \leq \frac{0.55-0.53}{0.009}) = P(-2.22 \leq Z \leq 2.22) = 0.9868 - 0.0132 = 0.9736$. (b) $P(V \geq 0.55) = P(Z \geq \frac{0.55-0.53}{0.009}) = P(Z \geq 2.22) = 1 - 0.9868 = 0.0132$.

**10.53:** (a) Because there are 10,000 equally likely four-digit numbers (0000 through 9999), the probability of an exact match is 1/10,000. (b) There is a total of 24 = 4(3)(2)(1) arrangements of the four digits 5, 9, 7, and 4 (there are four choices for the first digit, three for the second, two for the third), so the probability of a match in any order is 24/10,000.

**10.55:** (a)–(c) Results will vary, but after $n$ tosses, the distribution of the proportion $\hat{p}$ is approximately Normal with mean 0.5 and standard deviation $1/(2\sqrt{n})$, while the distribution of the count of heads is approximately Normal with mean $0.5n$ and standard deviation $\sqrt{n}/2$, so using the 68–95–99.7 rule, we have the results shown in the table on the right. Note that the range for $\hat{p}$ gets narrower, while the range for the count gets wider.

| $n$ | 99.7% Range for $\hat{p}$ | 99.7% Range for count |
|---|---|---|
| 40 | $0.5 \pm 0.237$ | $20 \pm 9.5$ |
| 120 | $0.5 \pm 0.137$ | $60 \pm 16.4$ |
| 240 | $0.5 \pm 0.097$ | $120 \pm 23.2$ |
| 480 | $0.5 \pm 0.068$ | $240 \pm 32.9$ |

**10.57:** (a) With $n = 20$, the variability in $\hat{p}$ is larger. With $n = 80$, nearly all answers will be between 0.24 and 0.56. With $n = 320$, nearly all answers will be between 0.32 and 0.48.

## Chapter 11: Sampling Distributions

**11.1:** Both 3.8 and 160.2 active cells per 100,000 cells are statistics (related to one sample—the subjects before infusion and the same subjects after infusion).

**11.3:** Both 12% and 23 are statistics, as they describe the sample of 230 American male weight lifters.

**11.5:** Although the probability of having to pay for a total loss for 1 or more of the 12 policies is very small, if this were to happen, it would be financially disastrous. On the other hand, for thousands of policies, the law of large numbers says that the average claim on many policies will be close to the mean, so the insurance company can be assured that the premiums they collect will (almost certainly) cover the claims.

**11.7 :** (a) Find the mean by summing the 10 observations and dividing by 10: $\mu = 694/10 = 69.4$.
(b) The table below shows the results for line 116. Note that we need to choose 5 digits because the digit 4 appears twice. (When choosing an SRS, no student should be chosen more than once.)
(c) The results for the other lines are in the table; the histogram is shown next to the table, though student histograms may vary in choice of intervals used. The center of the histogram is a bit lower than 69.4 (it is 66.9), but for a small group of sample means, we should not expect the center to be in exactly the right place.

| Line | Digits | Scores | $x$-bar |
|------|--------|--------|---------|
| 116 | 14459 | 63 + 72 + 72 + 59 = 266 | 66.5 |
| 117 | 3816 | 55 + 75 + 63 + 65 = 258 | 64.5 |
| 118 | 7319 | 66 + 55 + 63 + 59 = 243 | 60.75 |
| 119 | 95857 | 59 + 72 + 75 + 66 = 272 | 68 |
| 120 | 3547 | 55 + 72 + 72 + 66 = 265 | 66.25 |
| 121 | 7148 | 66 + 63 + 72 + 75 = 276 | 69 |
| 122 | 1387 | 63 + 55 + 75 + 66 = 259 | 64.75 |
| 123 | 54580 | 72 + 72 + 75 + 86 = 305 | 76.25 |
| 124 | 7103 | 66 + 63 + 86 + 55 = 270 | 67.5 |
| 125 | 9674 | 59 + 65 + 66 + 72 = 262 | 65.5 |

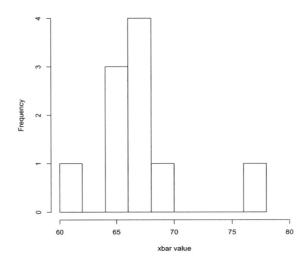

11.9 : (a) If $n = 100$, the sampling distribution of $\bar{x}$ is $N(186, 41/\sqrt{100}) = N(186$ mg/dl, 4.1 mg/dl). Therefore, $P(183 < \bar{x} < 189) = P\left(\dfrac{183-186}{4.1} < Z < \dfrac{189-186}{4.1}\right) = P(-0.73 < Z < 0.73) = 0.7673 - 0.2327 = 0.5346$. (b) With $n = 1000$, the sampling distribution of $\bar{x}$ is $N(186, 41/\sqrt{1000}) = N(186$ mg/dl, 1.2965 mg/dl) distribution, so $P(183 < \bar{x} < 189) = P\left(\dfrac{183-186}{1.2965} < Z < \dfrac{189-186}{1.2965}\right) P(-2.31 < Z < 2.31) = 0.9896 - 0.0104 = 0.9792$.

11.11: No: the histogram of the sample values will look like the population distribution, whatever it might happen to be. (For example, if we roll a fair die many times, the histogram of sample values should look relatively flat—probability close to 1/6 for each value 1, 2, 3, 4, 5, and 6.) The central limit theorem says that the histogram of *sample means* (from many large samples) will look more and more Normal.

11.13 : STATE: We ask what is the probability that the average loss for 10,000 such policies will be greater than $85, when the long-run average loss is $75? PLAN: Use the central limit theorem to approximate this probability. SOLVE: The central limit theorem says that, in spite of the skewness of the population distribution, the average loss among 10,000 policies will be approximately $N(\$75, \$300/\sqrt{10,000}) = N(\$75, \$3)$ Now $P(\bar{x} > \$85) = P\left(Z > \dfrac{85-75}{3}\right) = P(Z > 3.33) = 1 - 0.9996 = 0.0004$. CONCLUDE: We can be about 99.96% certain that average losses will not exceed $85 per policy.

11.15: (c) parameter. 58.8% is a proportion of all registered voters (the population).

11.17: (a) The mean of the sample means ($\bar{x}$'s) is the same as the population mean ($\mu$).

11.19: (a) "Unbiased" means that the estimator is right "on the average."

**11.21**: (b) For $n = 6$ women, $\bar{x}$ has a $N(266, 16/\sqrt{6}) = N(266, 6.5320)$ distribution, so $P(\bar{x} > 270) = P(Z > 0.61) = 0.2709$.

**11.23**: Both 25.40 and 20.41 are statistics (related, respectively, to the two samples).

**11.25**: $\bar{x}$ has mean $\mu = 852$ mm, and standard deviation $\sigma/\sqrt{n} = 82/\sqrt{10} = 25.93$ mm.

**11.27**: Let $X$ be Shelia's measured glucose level. (a) $P(X > 140) = P(Z > \frac{140-122}{12}) = P(Z > 1.5) = 1 - 0.9332 = 0.0668$. (b) If $\bar{x}$ is the mean of four measurements (assumed to be independent), then $\bar{x}$ has a $N(122, 12/\sqrt{4}) = N(122$ mg/dl, 6 mg/dl) distribution, and $P(\bar{x} > 140) = P(Z > \frac{140-122}{6}) = P(Z > 3) = 1 - 0.9987 = 0.0013$.

**11.29**: As shown in Exercise 11.27(b), the mean of four measurements has a $N(122$ mg/dl, 6 mg/dl) distribution, and $P(Z > 1.645) = 0.05$ if $Z$ is $N(0,1)$, so $L = 122 + (1.645)(6) = 131.87$ mg/dl.

**11.31**: (a) The central limit theorem gives that $\bar{x}$ will have a Normal distribution with mean 8.8 beats per five seconds, and standard deviation $1/\sqrt{12} = 0.288675$ beats per five seconds. (b) $P(\bar{x} < 8) = P(Z < -2.77) = 0.0028$. (c) If the total number of beats in one minute is less than 100, then the average over 12 5-seceond intervals needs to be less than $100/12 = 8.333$ beats per five seconds. $P(\bar{x} < 8.333) = P(Z < -1.62) = 0.0526$.

**11.33**: STATE: What are the probabilities of an average return over 10%, or less than 5%? PLAN: Use the central limit theorem to approximate this probability. SOLVE: The central limit theorem says that over 40 years, $\bar{x}$ (the mean return) is approximately Normal with mean $\mu = 10.8\%$ and standard deviation $17.1\%/\sqrt{40} = 2.704\%$. Therefore, $P(\bar{x} > 10\%) = P(Z > \frac{10-10.8}{2.704}) = P(Z > -0.30) = 0.6179$, and $P(\bar{x} < 5\%) = P(Z > \frac{5-10.8}{2.704})$ $P(Z < -2.14) = 0.0162$.
CONCLUDE: There is about a 62% chance of getting average returns over 10%, and a 1.6% chance of getting average returns less than 5%.

**Note**: *We have to assume that returns in separate years are independent. We also have to assume that the sample size (40 here) is large enough for the central limit theorem to hold. However, for virtually all populations a sample of 40 is large enough for this to hold.*

**11.35**: We need to choose $n$ so that $6.4/\sqrt{n} = 1$. That means $\sqrt{n} = 6.4$, so $n = 40.96$. Because $n$ must be a whole number, take $n = 41$.

**11.37**: On the average, Joe loses 40 cents each time he plays (that is, he spends $1 and gets back 60 cents).

**11.39**: (a) With $n = 150{,}000$, $\mu_{\bar{x}} = \$0.40$ and $\sigma_{\bar{x}} = \dfrac{\$18.96}{\sqrt{150{,}000}} = \$0.0490$. (b) $P(\$0.30 < \bar{x} < \$0.50) = P\left(\dfrac{0.30 - 0.40}{0.049} < Z < \dfrac{0.50 - 0.40}{0.049}\right) = P(-2.04 < Z < 2.04) = 0.9793 - 0.0207 = 0.9586$.

**11.41**: The mean is 10.5 (= (3)(3.5) because a single die has a mean of 3.5). Sketches will vary, as will the number of rolls; one result is shown.

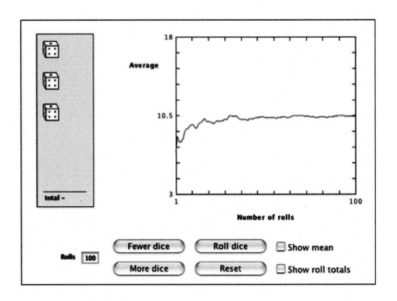

# Chapter 12: General Rules of Probability

**12.1:** It is unlikely that these events are independent. In particular, it is reasonable to expect that younger adults are more likely than older adults to be college students. **Note:** *Using the notation of conditional probability introduced later in this chapter, P(college student | over 55) < 0.08.*

**12.3:** If we assume that each site is independent of the others (and that they can be considered as a random sample from the collection of sites referenced in scientific journals), then $P$(all seven are still good) $= (0.87)^7 = 0.3773$.

**12.5:** (a) The entire area of event $A$ must be 0.70, the entire area of event $B$ must be 0.25, and the area at the intersection of A and B must be 0.05. This forces area 0.65 into the area in "$A$ and not $B$", and area 0.20 in the area "$B$ and not $A$". Since the total probability must be 1, the area outside $A$ and $B$ must be $1 - 0.65 - 0.05 - 0.20 = 0.10$. See the Venn diagram provided below.

(b) The events are:

{$A$ and $B$} = {student is at least 25 and local}
{$A$ and not $B$} = {student is at least 25 and not local}
{$B$ and not $A$} = {student is less than 25 and local}
{neither $A$ nor $B$} = {student is less than 25 and not local}

(c) $P(A$ and $B) = 0.05$, as indicated as the intersection of events $A$ and $B$. Subtracting this from the given probabilities for $A$ and $B$ gives $P(A$ and not $B)$ and $P(B$ and not $A)$ as 0.65 and 0.20, respectively. Those probabilities add to 0.90, so $P$(neither $A$ nor $B) = 0.10$.

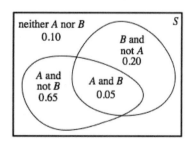

**12.7:** $P(B \mid \text{not } A) = \dfrac{P(B \text{ and not } A)}{P(\text{not } A)} = \dfrac{P(B) - P(B \text{ and } A)}{P(\text{not } A)} = \dfrac{0.2}{0.3} = 0.667$.

**12.9:** Let $H$ be the event that an adult belongs to a club, and $T$ be the event that he/she goes at least twice a week. We have been given $P(H) = 0.15$ and $P(T \mid H) = 0.50$. Note also that $P(T$ and $H) = P(T)$, since one has to be a member of the club in order to attend. So $P(T) = P(H)P(T \mid H) = (0.15)(0.50) = 0.075$. About 7.5% of all adults go to health clubs at least twice a week.

12.11: (a) and (b) These probabilities are provided in the table. Each time a spade is removed from the deck, we condition the next draw – one fewer spade remains in the deck, and one fewer card overall remains.

$$P(\text{1st card } \spadesuit) = \tfrac{13}{52} = \tfrac{1}{4} = 0.25$$
$$P(\text{2nd card } \spadesuit \mid 1 \spadesuit \text{ picked}) = \tfrac{12}{51} = \tfrac{4}{17} \doteq 0.2353$$
$$P(\text{3rd card } \spadesuit \mid 2 \spadesuit \text{s picked}) = \tfrac{11}{50} = 0.22$$
$$P(\text{4th card } \spadesuit \mid 3 \spadesuit \text{s picked}) = \tfrac{10}{49} \doteq 0.2041$$
$$P(\text{5th card } \spadesuit \mid 4 \spadesuit \text{s picked}) = \tfrac{9}{48} = \tfrac{3}{16} = 0.1875$$

(c) The product of these conditional probabilities gives the probability of a flush in spades by the general multiplication rule: We must draw a spade, and then another, and then a third, a fourth, and a fifth. The product of these probabilities is $0.25(0.2353)(0.22)(0.2041)(0.1875) = 0.000495$.
(d) Because there are four possible suits in which to have a flush, the probability of a flush is four times that found in (c), or about $4(0.000495) = 0.00198$.

12.13: STATE: What is the probability distribution for the number of people that have a peanut allergy in a random sample of 3 people? PLAN: The allergy status of individuals are independent events. Each person is a "success" with probability 0.01. We construct a tree diagram showing the results (allergic or not) for each of the three individuals. SOLVE: In the tree diagram, each "up-step" represents an allergic individual (and has probability 0.01), and each "down-step" is a non-allergic individual (and has probability 0.99). At the end of each of the 8 complete branches are the value of $X$. Any branch with 2 up-steps and 1 down-step has probability $(0.01)^2(0.99)^1 = 0.000099$, and yields $X = 2$. Any branch with 1 up-step and 2 down-steps has probability $(0.01)^1(0.99)^2 = 0.009801$, and yields $X = 1$.

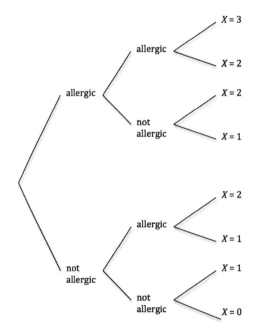

There are three branches each corresponding to $X = 2$ and $X = 1$, and only one branch each for $X = 3$ and $X = 0$. Because $X = 0$ and $X = 3$ appear on one branch each, $P(X = 0) = (0.99)^3 =$

0.970299 and $P(X = 3) = (0.01)^3 = 0.000001$. Meanwhile, $P(X=1) = 3(0.01)^1(0.99)^2 = 0.029403$, and $P(X=2)=3(0.01)^2(0.99)^1 = 0.000297$. CONCLUDE: $P(X = 0) = 0.970299$, $P(X = 1) = 0.029403$, $P(X = 2) = 0.000297$, and $P(X = 3) = 0.000001$.

12.15: With $X$ denoting the number of people in our random sample with a peanut allergy,
$$P(X = 2 | X \geq 1) = \frac{P(X = 2 \text{ and } X \geq 1)}{P(X \geq 1)} = \frac{P(X = 2)}{P(X \geq 1)} = \frac{0.000297}{1 - 0.970299} = 0.010.$$

(Note that the event $\{X = 2 \text{ and } X \geq 1\}$ is the same as the event $\{X = 2\}$.)

12.17: (b) This probability is $(0.98)^3 = 0.9412$.

12.19: (a) $P$(at least one positive) $= 1 - P$(both negative) $= 1 - P$(first negative)$P$(second negative) $= 1 - (0.1)(0.2) = 0.98$.

12.21: (c) Of $6,006 + 1,623 = 7,629$ Engineering doctorates, 1,623 were awarded to females. Then P(female | engineering) $= 1,623/7,629 = 02127$, or 0.21.

12.23: (c) We want the fraction of engineering doctorates conferred to women. Hence, $A$ (engineering degree) is what has been given. Hence, $P(B | A)$.

12.25: (c) $P(W \text{ and } D) = P(W)P(D | W)=(0.86)(0.028) = 0.024$.

12.27: $P(8 \text{ losses}) = (0.75)^8 = 0.1001$.

12.29: (a) $P$(win the jackpot) $= \left(\frac{1}{20}\right)\left(\frac{9}{20}\right)\left(\frac{1}{20}\right) = 0.001125$. (b) The other (non-cherry) symbol can show up on the middle wheel, with probability $\left(\frac{1}{20}\right)\left(\frac{11}{20}\right)\left(\frac{1}{20}\right) = 0.001375$, or on either of the outside wheels, with probability $= \left(\frac{19}{20}\right)\left(\frac{9}{20}\right)\left(\frac{1}{20}\right)$ (each). (c) Combining all three cases from part (b), we have $P$(exactly two cherries) $= 0.001375 + 2(0.021375) = 0.044125$.

12.31: STATE: What proportion of operations succed and are free from infection? PLAN: Let $I$ be the event "infection occurs" and let $F$ be "the repair fails." We have been given $P(I) = 0.03$, $P(F) = 0.14$, and $P(I \text{ and } F) = 0.01$. We want to find $P(\text{not } I \text{ and not } F)$. SOLVE: First use the general addition rule: $P(I \text{ or } F) = P(I) + P(F) - P(I \text{ and } F) = 0.03 + 0.14 - 0.01 = 0.16$. This is the shaded region in the Venn diagram provided. Now observe that the desired probability is the complement of "$I$ or $F$" (the *unshaded* region): $P(\text{not } I \text{ and not } F) = 1 - P(I \text{ or } F) = 0.84$. CONCLUDE: 84% of operations succeed and are free from infection.

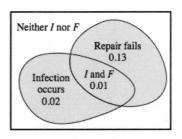

12.33: PLAN: Let $I$ be the event "infection occurs" and let $F$ be "the repair fails." Refer to the Venn diagram in Exercise 12.31 (ignoring the shading). We want to find $P(I \mid \text{not } F)$. SOLVE: We have $P(I \mid \text{not } F) = \dfrac{P(I \text{ and not } F)}{P(\text{not } F)} = \dfrac{0.2}{0.86} = 0.0233$. CONCLUDE: The probability of infection given that the repair is successful is 0.0233. That is, in 2.33% of all successful operation cases, the patient develops infection.

12.35: Let $H$ be the event student was home schooled. Let $R$ be the event student attended a regular public school. We want $P(H \mid \text{not } R)$. Note that the event "$H$ and not $R$" = "$H$", since the events are disjoint. Then $P(H \mid \text{not } R) = \dfrac{P(H)}{P(\text{not } R)} = \dfrac{0.006}{1 - 0.781} = 0.0274$.

12.37: (a) These events are not independent, because $P$(pizza with mushrooms) = 4/7, but $P$(mushrooms | thick crust) = 2/3 (if the events were independent, these probabilities would be equal). Alternatively, note that $P$(thick crust with mushrooms) = 2/7, which is not equal to the product of $P$(mushrooms) = 4/7 and $P$(thick crust pizza) = 3/7. (b) With the eighth pizza, $P$(mushrooms) = 4/8 = 1/2, and $P$ (mushrooms | thick crust) = 2/4 = 1/2, so these events are independent.

12.39: Let $W$ be the event "the person is a woman" and $M$ be "the person earned a Master's degree." (a) $P(\text{not } W)$ = 1421/3560 = 0.3992. (b) $P(\text{not } W \mid M)$ = 282/732 = 0.3852. (c) The events "choose a man" and "choose a Master's degree recipient" are not independent. If they were, the two probabilities in (a) and (b) would be equal.

12.41: Let $D$ be the event "a seedling was damaged by a deer." (a) $P(D)$ = 209/871 = 0.2400. (b) The conditional probabilities are:

$P(D \mid \text{no cover})$ = 60/211 = 0.2844

$P(D \mid \text{cover} < 1/3)$ = 76/234 = 0.3248

$P(D \mid 1/3 \text{ to } 2/3 \text{ cover})$ = 44/221 = 0.1991

$P(D \mid \text{cover} > 2/3)$ = 29/205 = 0.1415

(c) Cover and damage are not independent; $P(D)$ decreases noticeably when thorny cover is 1/3 or more.

12.43: This conditional probability is $P(\text{cover} < 1/3 \mid D)$ = 76/(60 + 76 + 44 + 29) = 76/209 = 0.3636, or 36.36%.

12.45: This is $P(A \text{ and not } B \text{ and not } C)$ = 0.35.

**12.47:** (a) P(doubles on first toss) = 1/6, since 6 of the 36 equally likely outcomes enumerated in Figure 10.2 involve rolling doubles. (b) We need no doubles on the first roll (which happens with probability 5/6), then doubles on the second toss. P(first doubles appears on toss 2) = (5/6)(1/6) = 5/36. (c) Similarly, P(first doubles appears on toss 3) = $(5/6)^2(1/6)$ = 25/216. (d) P(first doubles appears on toss 4) = $(5/6)^3(1/6)$, etc. In general, P(first doubles appears on toss k) = $(5/6)^{k-1}(1/6)$. (e) P(go again within 3 turns) = P(roll doubles in 3 or fewer rolls) = P(roll doubles on 1st, 2nd or 3rd try) = (1/6) + (5/6)(1/6) + $(5/6)^2(1/6)$ = 0.4213.

**12.49:** STATE: What percentage of the vote should the black candidate in this election expect? PLAN: Let W, B, and H be the events that a randomly selected voter is (respectively) white, black, and Hispanic. We have been given P(W) = 0.4, P(B) = 0.4, P(H) = 0.2. If F = "a voter votes for the candidate," then P(F | W) = 0.3, P(F | B) = 0.9, P(F | H) = 0.5. We want to find P(F). SOLVE: The tree diagram provided organizes the information. The numbers on the right side of the tree are found by the general multiplication rule; for example, P("white" and "for") = P(W and F) = P(W) P(F | W) = (0.4)(0.3) = 0.12. We find P(F) by adding all the numbers next to the branches ending in "for": P(F) = 0.12 + 0.36 + 0.10 = 0.58. CONCLUDE: The black candidate expects to get 58% of the vote.

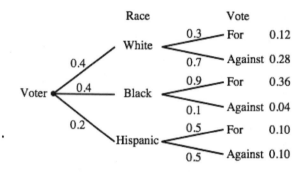

**12.51:** $P(B \mid F) = \dfrac{P(B \text{ and } F)}{P(F)} = \dfrac{0.36}{0.58} = 0.6207$, or about 62%.

**12.53:** For a randomly selected resident of the United States, let W, B, A, and L be (respectively) the events that this person is white, black, Asian, and lactose intolerant. We have been given

$P(W) = 0.82 \qquad P(B) = 0.14 \qquad P(A) = 0.04$

$P(L|W) = 0.15 \qquad P(L|B) = 0.70 \qquad P(L|A) = 0.90$

(a) P(L) = (0.82)(0.15) + (0.14)(0.70) + (0.04)(0.90) = 0.257, or 25.7%. (b) P(A | L) = P(A and L)/P(L) = (0.04)(0.90)/0.257 = 0.1401, or 14%.

**12.55:** In this problem, allele 29 is playing the role of A, and 0.181 is the proportion with this allele (a = 0.181). Similarly, allele 31 is playing the role of B, and the proportion having this allele is b = 0.071. The labels aren't important — you can reverse assignments of A and B. The proportion of the population with combination (29,31) is therefore 2(0.181)(0.071) = 0.025702. The proportion with combination (29,29) is (0.181)(0.181) = 0.032761. Of course under these assignments, there are other alleles possible for this locus.

12.57: In Exercise 12.55, we found that the proportion of the population with allele (29,31) at loci D21S11 is 0.025702. In Exercise 12.56, we found that the proportion with allele (16,17) at loci D3S1358 is 0.098368. Assuming independence between loci, the proportion with allele (29,31) at D21S11 and (16,17) at D3S1358 is (0.098368)(0.025702) = 0.002529.

## Chapter 13: Binomial Distributions

13.1: Binomial. (1) We have a fixed number of observations ($n = 15$). (2) It is reasonable to believe that each call is independent of the others. (3) "Success" means reaching a live person, "failure" is any other outcome. (4) Each randomly dialed number has chance $p = 0.2$ of reaching a live person.

13.3: Not binomial. The trials aren't independent. If one tile in a box is cracked, there are likely more tiles cracked.

13.5: (a) $C$, the number caught, is binomial with $n = 10$ and $p = 0.7$. $M$, the number missed, is binomial with $n = 10$ and $p = 0.3$. (b) We find $P(M = 3) = \binom{10}{3}(0.3)^3(0.7)^7 = (120)(0.027)(0.08235) = 0.2668$. With software, we find $P(M \geq 3) = 0.6172$.

13.7: The screenshots below show Google's answers at the time these solutions were prepared. (a) (5 choose 2) returns 10. (b) (500 choose 2) returns the answer 124,750, and (500 choose 100) returns the answer $(2.04169424)(10)^{107}$. (c) (10 choose 1)*0.1*0.9^9 returns 0.387420489.

13.9: (a) $X$ is binomial with $n = 10$ and $p = 0.3$; $Y$ is binomial with $n = 10$ and $p = 0.7$ (b) The mean of $Y$ is $(10)(0.7) = 7$ errors caught, and for $X$ the mean is $(10)(0.3) = 3$ errors missed. (c) The standard deviation of $Y$ (or $X$) is $\sigma = \sqrt{10(0.7)(0.3)} = 1.4491$ errors.

13.11: Let $X$ be the number of admitted students that decide to attend. Then $X$ has the Binomial distribution with $n = 1520$ and $p = 0.31$. (a) $\mu = np = (1520)(0.31) = 471.2$ and $\sigma = \sqrt{np(1-p)} = \sqrt{1520(0.31)(1-0.31)} = \sqrt{325.128} = 18.0313$ students. (b) Note that $np = (1520)(0.31) = 471.2 \geq 10$ and $n(1-p) = (1520)(0.69) = 1048.8 \geq 10$, so $n$ is large enough for the Normal approximation to be reasonable. The college wants 475 students, so more than the number needed is 476 or more. Then $P(X \geq 476) = P\left(Z \geq \dfrac{476 - 471.2}{18.0313}\right) = P(Z \geq 0.27) = 0.3936$.

(c) The exact probability is 0.4045 (obtained from software), so the Normal approximation is 0.0109 too low. For a better approximation, consider using the continuity correction, described in later exercises.

13.13: (b) He has 3 independent eggs, each with probability 1/4 of containing salmonella.

13.15: (c) The selections are not independent; once we choose one student, it changes the probability that the next student is a business major.

13.17: (a) This probability is $(0.60)^2(0.40)^3 = 0.02304$.

13.19: (b) This is the event that a single digit is 8 or 9, so the probability is 0.20.

13.21: (a) The mean is $np = (80)(0.20) = 16$.

13.23: (a) A binomial distribution is *not* an appropriate choice for field goals made, because given the different situations the kicker faces, his probability of success is likely to change from one attempt to another. (b) It would be reasonable to use a binomial distribution for free throws made because we have $n = 150$ attempts, presumably independent (or at least approximately so), with chance of success $p = 0.8$ each time.

13.25: (a) $X$ has the binomial distribution with $n = 5$ and $p = 0.65$. (b) The possible values of $X$ are the integers 0, 1, 2, 3, 4, 5. (c) All cases are computed:

$$P(X=0) = \binom{5}{0}(0.65)^0(0.35)^5 = 0.00525 \qquad P(X=1) = \binom{5}{1}(0.65)^1(0.35)^4 = 0.04877$$

$$P(X=2) = \binom{5}{2}(0.65)^2(0.35)^3 = 0.18115 \qquad P(X=3) = \binom{5}{3}(0.65)^3(0.35)^2 = 0.33642$$

$$P(X=4) = \binom{5}{4}(0.65)^4(0.35)^1 = 0.31239 \qquad P(X=5) = \binom{5}{5}(0.65)^5(0.35)^0 = 0.11603.$$

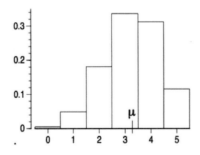

(d) $\mu = np = (5)(0.65) = 3.25$ and $\sigma = \sqrt{5(0.65)(1-0.65)} = 1.0665$ years. The mean $\mu$ is indicated on the probability histogram.

13.27: (a) All women are independent, and each has the same probability of getting pregnant. (b) Under ideal conditions, the number who get pregnant is binomial with $n = 20$ and $p = 0.01$; $P(N \geq 1) = 1 - P(N = 0) = 1 - 0.8179 = 0.1821$. In typical use, $p = 0.03$, and $P(N \geq 1) = 1 - 0.5438 = 0.4562$.

13.29: (a) $X$, the number of women who get pregnant in typical use, is binomial with $n = 600$ and $p = 0.03$. The Normal approximation is safe: $np = 18$ and $n(1 - p) = 582$ are both larger than 10. The mean is 18 and the standard deviation is 4.1785, so $P(X \geq 20) = P\left(Z \geq \dfrac{20 - 18}{4.1785}\right) = P(Z \geq 0.48) = 0.3156$. The exact binomial probability is 0.3477. (b) Under ideal conditions, $p = 0.01$, so $np = 6$ is too small.

13.31: (a) If $R$ is the number of red-blossomed plants out of a sample of 4, then $P(R = 3) = \binom{4}{3}(0.75)^3(0.25)^1 = 0.4219$, using a binomial distribution with $n = 4$ and $p = 0.75$. (b) With $n = 60$, the mean number of red-blossomed plants is $np = 45$. (c) If $R$ is the number of red-blossomed plants out of a sample of 60, then $P(R \geq 45) = P(Z \geq 0) = 0.5000$ (software gives 0.5688 using the binomial distribution).

13.33: (a) Of 1,498,000 total vehicles in these top 5 nameplates, Impalas accounted for proportion $184{,}000/1{,}498{,}000 = 0.12283$. (b) If $I$ is the number of Impala buyers in the 1000 surveyed buyers, then $I$ has the binomial distribution with $n = 1000$, and $p = 0.12283$. Hence, $\mu = np = (1000)(0.12283) = 122.83$ and $\sigma = \sqrt{np(1-p)} = \sqrt{1000(0.12283)(1 - 0.12283)} = 10.38$ Impala buyers. (c) $P(I > 100) = P(I \geq 101) = P(Z \geq -2.10) = 0.9821$.

13.35: (a) With $n = 100$, the mean and standard deviation are $\mu = 75$ and $\sigma = 4.3301$ questions, so $P(70 \leq X \leq 80) = P(-1.15 \leq Z \leq 1.15) = 0.7498$ (software gives 0.7967). (b) With $n = 250$, we have $\mu = 187.5$ and $\sigma = 6.8465$ questions, and a score between 70% and 80% means 175 to 200 correct answers, so $P(175 \leq X \leq 200) = P(-1.83 \leq Z \leq 1.83) = 0.9328$ (software gives 0.9428).

13.37: (a) Answers will vary, but over 99.8% of samples should have 0 to 4 bad CDs. (b) Each time we choose a sample of size 10, the probability that we have exactly 1 bad CD is 0.3874; therefore, out of 20 samples, the number of times that we have exactly 1 bad CD has a binomial distribution with parameters $n = 20$ and $p = 0.3874$. This means that most students—99.8% of them—will find that between 2 and 14 of their 20 samples have exactly 1 bad CD, giving a proportion between 0.10 and 0.70. (If anyone has an answer outside of that range, which would be significant evidence that he or she did the exercise incorrectly.)

13.39: The number $N$ of infections among untreated BJU students is binomial with $n = 1400$ and $p = 0.80$, so the mean is 1120 and the standard deviation is 14.9666 students. 75% of that group is 1050, and the Normal approximation is safe: $P(N \geq 1050) = P\left(Z \geq \dfrac{1050 - 1120}{14.9666}\right) = P(Z \geq -4.68)$, which is very near to 1. (Exact computation gives 0.999998.)

13.41: Let $V$ and $U$ be (respectively) the number of new infections among the vaccinated and unvaccinated children. $V$ is binomial with $n = 17$ and $p = 0.05$, with mean 0.85 infections. $U$ is binomial with $n = 3$ and $p = 0.80$, with mean 2.4 infections. (a) $P(V = 1) = 0.3741$ and $P(U = 1) = 0.0960$. Because these events are independent, $P(V = 1 \text{ and } U = 1) = P(V = 1)P(U = 1) = 0.0359$. (b) Considering all the possible ways to have a total of 2 infections, we have $P(2 \text{ infections}) = P(V = 0 \text{ and } U = 2) + P(V = 1 \text{ and } U = 1) + P(V = 2 \text{ and } U = 0) = P(V = 0)P(U = 2) + P(V = 1)P(U = 1) + P(V = 2)P(U = 0) = (0.4181)(0.3840) + (0.3741)(0.0960) + (0.1575)(0.0080) = 0.1977$.

13.43: The number $X$ of fairways Phil hits is binomial with $n = 24$ and $p = 0.52$. (a) Since $np = 12.48$ and $n(1 - p) = 11.52$ are both larger than 10 (barely), the Normal approximation is (barely) safe. (b) The mean is $np = 12.48$ and the standard deviation is $\sqrt{np(1 - p)} = \sqrt{24(0.52)(0.48)} = 2.447529$. Using the Normal approximation, $P(X \geq 17) = P(Z \geq \frac{17 - 12.48}{2.447529}) = P(Z \geq 1.85) = 1 - 0.9678 = 0.0322$. (c) With the continuity correction, $P(X \geq 17) = P(X \geq 16.5) = P(Z \geq \frac{16.5 - 12.48}{2.447529}) = P(Z \geq 1.64) = 1 - 0.9495 = 0.0505$. Indeed, the answer using the continuity correction is closer to the exact answer (0.0487, using software).

# Chapter 14: Confidence Intervals: The Basics

14.1: (a) The sampling distribution of $\bar{x}$ has mean $\mu$ (unknown) and standard deviation $\dfrac{\sigma}{\sqrt{n}}$ $= \dfrac{34}{\sqrt{51{,}000}} = 0.1506$. (b) According to this rule, 95% of all values of $\bar{x}$ fall within 2 standard deviations of the sampling distribution of $\mu$. That is, within $2(0.1506) = 0.3012$. (c) $153 \pm 0.3012$, or between 152.7 and 153.3.

14.3: Shown below are sample output screens for (a) 10 and (b) 1000 SRS's. In 99.4% of all repetitions of part (a), students should see between 5 and 10 hits (that is, at least 5 of the 10 SRS's capture the true mean $\mu$). Out of 1000 80% confidence intervals, nearly all students will observe between 76% and 84% capturing the mean.

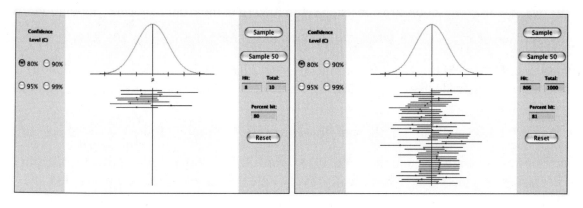

14.5: Search Table A for 0.075 (half of the 15% that is not included in the middle, shaded area corresponding to 85% confidence). This area corresponds to $-z^* = -1.44$, or $z^* = 1.44$.

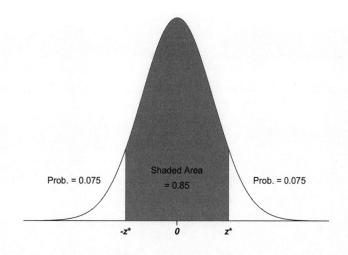

14.7: (a) A stemplot is provided. The two low scores (72 and 74) are both possible outliers, but there are no other apparent deviations from Normality.

```
 7 | 24
 7 |
 8 |
 8 | 69
 9 | 13
 9 | 68
10 | 023334
10 | 578
11 | 11222444
11 | 89
12 | 0
12 | 8
13 | 02
```

(b) STATE: What is the mean IQ $\mu$ of all seventh-grade girls in this school district?  PLAN: We will estimate $\mu$ by giving a 99% confidence interval.  SOLVE: The problem states that these girls are an SRS of the population, which is very large, so conditions for inference are met.  In part (a), we saw that the scores are consistent with having come from a Normal population.  With $\bar{x} = 105.84$, and $z^* = 2.576$, our 99% confidence interval for $\mu$ is given by $105.84 \pm 2.576 \frac{15}{\sqrt{31}}$ $= 105.84 \pm 6.94 = 98.90$ to $112.78$ IQ points.  CONCLUDE: We are 99% confident that the mean IQ of seventh-grade girls in this district is between 98.90 and 112.78 points.

14.9: With $z^* = 1.96$ and $\sigma = 7.5$, the margin of error is $z^* \sigma/\sqrt{n} = 1.96 \frac{7.5}{\sqrt{n}} = \frac{14.7}{\sqrt{n}}$.  (a) and (b) The margins of error are given in the table, using $n = 100$, $n = 400$ and $n = 1600$.  (c) Margin of error decreases as $n$ increases.  (Specifically, every time the sample size $n$ is quadrupled, the margin of error is halved.)

| $n$ | m.e. |
|---|---|
| 100 | 1.47 |
| 400 | 0.735 |
| 1600 | 0.3675 |

14.11: (c) $z = 3.291$.  Using Table A, search for 0.9995.

14.13: (b) As the confidence level increases, $z^*$ increases.  This makes the margin of error larger.

14.15: (b) The standard deviation of $\bar{x}$ is $\frac{\sigma}{\sqrt{n}} = \frac{35}{\sqrt{900}} = 1.167$.

14.17: (b) As the confidence level increases, $z^*$ increases.  This makes the margin of error larger.

14.19: (a) We use $\bar{x} \pm z^* \frac{\sigma}{\sqrt{n}}$, or $118 \pm 2.576 \frac{65}{\sqrt{463}} = 118 \pm 7.78 = 110.22$ to $125.78$ minutes.
(b) The 463 students in this class must be a random sample of all of the first-year students at this university to satisfy conditions for inference.

14.21: The margin of error is now $2.576 \frac{65}{\sqrt{464}} = 7.77$, so the extra observation has minimal impact on the margin of error (the sample was large to begin with). If $\bar{x} = 247$, then the 99% confidence interval for average amount of time spent studying becomes $247 \pm 7.77 = 239.23$ to 254.77 minutes. The outlier had a huge impact on $\bar{x}$, which shifts the interval a lot.

14.23: This student is also confused. If we repeated the sample over and over, 95% of all future sample means would be within 1.96 standard deviations of $\mu$ (that is, within $1.96 \frac{\sigma}{\sqrt{n}}$) of the true, unknown value of $\mu$. Future samples will have no memory of our sample.

14.25: (a) A stemplot of the data is provided. Notice that the distribution is noticeably skewed to the left. The data do not appear to follow a Normal distribution.

```
23 | 0
24 | 0
25 |
26 | 5
27 |
28 | 7
29 |
30 | 149
31 | 389
32 | 033577
33 | 0126
```

(b) STATE: What is the mean load $\mu$ required to pull apart pieces of Douglas fir? PLAN: We will estimate $\mu$ by giving a 95% confidence interval. SOLVE: The problem states that we are willing to take this sample to be an SRS of the population. In spite of the shape of the stemplot, we are told to assume that this distribution is Normal with standard deviation $\sigma = 3000$ lb. We find $\bar{x} = 30{,}841$ lb, so the 95% confidence interval for $\mu$ is given by $30{,}841 \pm 1.96 \frac{3000}{\sqrt{20}} = 30{,}841 \pm 1314.81 = 29{,}526.19$ to $32{,}155.81$ pounds. CONCLUDE: With 95% confidence, the mean load $\mu$ required to break apart pieces of Douglas fir is between 29,526.2 and 32,155.8 pounds.

14.27: (a) A stemplot is given. There is little evidence that the sample does not come from a Normal distribution. For inference, we must assume that the 10 untrained students were selected randomly from the population of all untrained people.

```
1  9
2  2 3 9
3  0 0 1 3 5
4  2
```

(b) STATE: What is the average (mean) DMS odor threshold, $\mu$, for all untrained people? PLAN: We will estimate $\mu$ with a 95% confidence interval. SOLVE: We have assumed that we have a random sample, and that the population we're sampling from is Normal. We obtain $\bar{x} = 29.4$ $\mu g/l$. Our 95% confidence interval for $\mu$ is given by $29.4 \pm 1.96 \frac{7}{\sqrt{10}}$ $(7)/\sqrt{10} = 29.4 \pm 4.34 = 25.06$ to $33.74$ $\mu g/l$. CONCLUDE: With 95% confidence, the mean sensitivity for all untrained people is between 25.06 and 33.74 $\mu g/l$..

# Chapter 15: Tests of Significance: The Basics

**15.1:** If $\mu = 115$ and $\sigma = 6$, the sampling distribution of the sample mean based on $n$ observations is approximately Normal with mean $\mu = 115$ and standard deviation $\frac{\sigma}{\sqrt{n}} = 6$. (b) The actual result lies out toward the high tail of the curve, while 118.6 is fairly close to the middle. If $\mu = 115$, observing a value similar to 118.6 would not be too surprising, but 125.8 is less likely, and it therefore provides some evidence that $\mu > 115$.

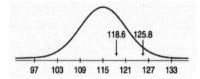

**15.3:** $H_0: \mu = 115$ vs. $H_a: \mu > 115$. Because the teacher suspects that older students have a higher mean, we have a one-sided alternative.

**15.5:** $H_0: \mu = 75$ vs. $H_a: \mu < 75$. The professor suspects this TA's students perform worse than the population of all students in the class on average.

**15.7:** Hypotheses are statements about parameters, not statistics. The research question is not about the sample mean ($\bar{x}$), but should be about the population mean, $\mu$.

**15.9:** With $\sigma = 60$ and $n = 18$, the standard deviation is $\frac{\sigma}{\sqrt{18}} = 14.1421$, so when $\mu = 0$, the distribution of $\bar{x}$ is $N(0, 14.1421)$. (b) The P-value is $P = 2P(\bar{x} \geq 17) = 2P\left(Z \geq \frac{17-0}{14.1421}\right) = 0.2302$.

**15.11:** (a) Using the Applet, the P-value for $\bar{x} = 118.6$ is 0.2743. This is not significant at either $\alpha = 0.05$ or $\alpha = 0.01$, since the P-value is greater than either value of $\alpha$. (b) The P-value for $\bar{x} = 125.8$ is 0.0359. This is significant at $\alpha = 0.05$, but not at $\alpha = 0.01$, since the P-value is less than 0.05 but greater than 0.01. (c) If $\mu = 115$ (that is, if $H_0$ were true), observing a value similar to 118.6 would not be too surprising, but 125.8 is less likely, and it therefore provides fairly strong evidence that $\mu > 115$.

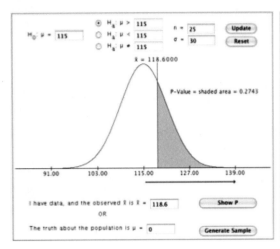
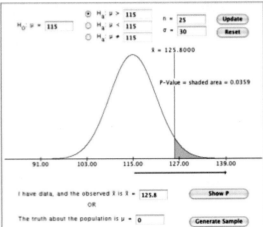

15.13: (a) $z = \dfrac{0.3 - 0}{1/\sqrt{10}} = \dfrac{0.3 - 0}{0.3162} = 0.9488$.  (b) $z = \dfrac{1.02 - 0}{1/\sqrt{10}} = \dfrac{1.02 - 0}{0.3162} = 3.226$.

(c) $z = \dfrac{17 - 0}{60/\sqrt{18}} = \dfrac{17 - 0}{14.1421} = 1.2021$. Note that in (c) the test is two-sided, while in (a) and (b) it is one-sided.

15.15: STATE: Is there evidence that the average percent tip when bad news is received (a bad weather prediction) is less than 20%? PLAN: Let $\mu$ be the average percentage tip for all customers receiving bad news. We test $H_0: \mu = 20$ against $H_a: \mu < 20$ since we wonder if the value of $\mu$ is less than 20%. SOLVE: We have a sample of $n = 20$ customers, and observe $\bar{x} = 18.19\%$. The standard deviation of $\bar{x}$ is $\dfrac{2}{\sqrt{20}} = 0.4472$, so the test statistic is $z = \dfrac{17.69 - 20}{0.4472} = -4.05$. The $P$-value is $P(Z \le -5.1655) \approx 0$. CONCLUDE: There is overwhelming evidence that the average tip percentage when bad news is delivered is lower than the average tip percentage overall. Random chance does not explain the small value of $\bar{x}$ observed.

15.17: This is not significant at the $\alpha = 0.05$ level because $z$ is not larger than 1.96 or less than $-1.96$. It is not significant at the $\alpha = 0.01$ level since $z$ is smaller than 2.576.

15.19: (a) This is the definition of a $P$-value.

15.21: (c) The $P$-value for $z = 2.433$ is 0.0075 (assuming that the difference is in the correct direction; that is, assuming that the alternative hypothesis was $H_a: \mu > \mu_0$).

15.23: (a) The null hypothesis states that $\mu$ takes on the "default" value, 18 seconds.

15.25: (c) The $P$-value refers to the probability of getting a sample as contrary to the null hypothesis as the sample observed, assuming $H_0$ is true.

15.27: (b) This is a one-sided alternative, so we have 0.005 in the right tail of the Normal distribution, leading to $z > 2.807$.

**15.29:** (a) We test $H_0: \mu = 0$ vs. $H_a: \mu > 0$. (b) $z = \dfrac{2.35 - 0}{2.5/\sqrt{200}} = 13.29$. (c) This value of $z$ is far outside the range we would expect from the $N(0,1)$ distribution. Under $H_0$, it would be virtually impossible to observe a sample mean as large as 2.35 based on a sample of 200 men. Hence, the sample mean is not explained by random chance, and we would easily reject $H_0$.

**15.31:** "$P = 0.03$" *does* mean that $H_0$ is not likely to be correct... but only in the sense that it provides a poor explanation of the data observed. It means that if $H_0$ is true, a sample as contrary to $H_0$ as our sample would occur by chance alone only 3% of the time if the experiment was repeated over and over. However, it does *not* mean that there is a 3% chance that $H_0$ is true.

**15.33:** The person making the objection is confusing practical significance with statistical significance. In fact, a 5% increase isn't a lot in a pragmatic sense. However, $P = 0.03$ means that random chance does not easily explain the difference observed. That is, there does seem to be an increase in mean improvement for those that expressed their anxieties... but the significance test does not address whether the difference is large enough to matter. Statistical significance is not practical significance.

**15.35:** In the sketch, the "significant at 1%" region includes only the dark shading ($z > 2.326$). The "significant at 5%" region of the sketch includes both the light and dark shading ($z > 1.645$). When a test is significant at the 1% level, it means that if the null hypothesis were true, outcomes similar to (or more extreme than) those seen are expected less than once in 100 repetitions of the experiment. When a test is significant at the 5% level, it means that if the null hypothesis were true, outcomes similar to (or more extreme than) those seen are expected less than five in 100 repetitions of the experiment. Hence, significance at the 1% level implies significance at the 5% level (or at any level higher than 1%). The converse is false: something that occurs "less than 5 times in 100 repetitions" is not necessarily as rare as something that happens "less than once in 100 repetitions," so a test that is significant at the 5% level is not necessarily significant at the 1% level.

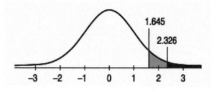

**15.37:** (a) Because a $P$-value is a probability, it can never be greater than 1. (b) The correct $P$-value is $P(Z \geq 1.33) = 0.0918$.

**15.39:** STATE: What is the mean percent change $\mu$ in spinal mineral content of nursing mothers? PLAN: We will test the hypotheses $H_0: \mu = 0\%$ against $H_a: \mu < 0\%$. SOLVE: With a sample of 47 observations, conditions required for use of the one-sample $z$-test are met. The sample mean is $\bar{x} = -3.587\%$. The test statistic is $z = \dfrac{-3.857 - 0}{2.5/\sqrt{47}} = -9.84$, and the $P$-value is $P(Z \leq -9.84) \approx 0$. CONCLUDE: There is overwhelming evidence that, on average, nursing mothers lose bone mineral.

15.41: (a) We test $H_0 : \mu = 0$ vs. $H_a : \mu > 0$, where $\mu$ is the mean sensitivity difference in the population. (b) STATE: Does eye grease have a significant impact on eye sensitivity? PLAN: We test the hypotheses $H_0 : \mu = 0$ vs. $H_a : \mu > 0$, as explained above. SOLVE: The mean of the 16 differences is $\bar{x} = 0.10125$, so the test statistic is $z = \dfrac{0.10125 - 0}{0.22/\sqrt{16}} = 1.84$. The one-sided P-value for this value of $z$ is $P = 0.0329$. CONCLUDE: The sample gives fairly strong evidence (at the $\alpha = 0.05$ level) that eye grease increases eye sensitivity.

15.43: (a) No, because 33 falls in the 95% confidence interval, which is (27.5, 33.9). (b) Yes, because 34 does not fall in the 95% confidence interval.

# Chapter 16: Inference in Practice

**16.1:** The most important reason is (c); this is a convenience sample consisting of the first 20 students on a list. This is not an SRS. Anything we learn from this sample will not extend to the larger population. The other two reasons are valid, but less important issues. Reason (a) — the size of the sample and large margin of error — would make the interval less informative, even if the sample were representative of the population. Reason (b)—nonresponse—is a potential problem with every survey, but there is no particular reason to believe it is more likely in this situation.

**16.3:** Any number of things could go wrong with this convenience sample. The day after Thanksgiving is widely regarded (rightly or wrongly) as a day on which retailers offer great deals—and the kinds of shoppers found that day probably don't represent shoppers generally. Also, the sample isn't random.

**16.5:** No. The confidence interval does not describe the range of future values of $\bar{x}$. Instead, if we repeated the experiment over and over, each time computing a 95% confidence interval for the population mean $\mu$, then 95% of these confidence intervals would capture that the true population mean.

**16.7:** The margin of error only addresses chance variation in the random selection of a sample. Hence, the answer is (c). Sources of bias described in (a) and (b) are not accounted for in the margin of error, and are difficult to assess.

**16.9:** The full applet output for $n = 5$ is below on the left. Next to this are the Normal curves drawn for $n = 15$ and $n = 40$. The reported P-values agree with the "hand-computed" values $z = \dfrac{4.8-5}{0.5/\sqrt{n}}$, and $P = P(Z \leq z)$, given in the table. For example, if $n = 40$, $\sigma_{\bar{x}} = \dfrac{0.5}{\sqrt{40}} = 0.0791$, so the P-value $= P\left(Z \leq \dfrac{4.8-5}{0.0791}\right) = P(Z \leq -2.53) = 1 - 0.9943 = 0.0057$.

| $n$ | $z$ | $P$ |
|---|---|---|
| 5 | −0.89 | 0.1867 |
| 15 | −1.55 | 0.0606 |
| 40 | −2.53 | 0.0057 |

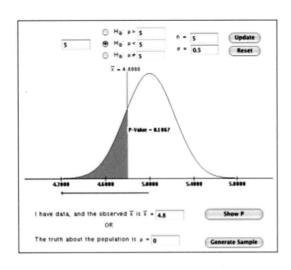
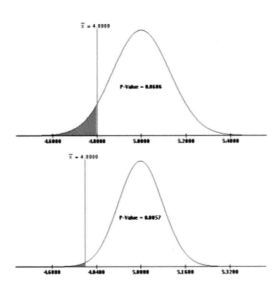

**16.11:** (a) In a sample of size $n = 500$, we expect to see about 5 people who have a P-value of 0.01 or less [ $5 = (500)(0.01)$ ]. These 4 might have ESP, or they may simply be among the "lucky" ones that occurred by random chance, as expected. (b) The researcher should repeat the procedure on these 4 subjects to see whether they again perform well.

**16.13:** We desire to know the sample size needed to estimate the mean to within margin of error 10 with 90% confidence. We know that $\sigma = 30$ and $z^* = 1.645$, so the sample size required is $n = \left(\dfrac{z^*\sigma}{m}\right)^2 = \left(\dfrac{(1.645)(30)}{10}\right)^2 = 24.354$. We always round up, here to $n = 25$. A sample of 25 is required.

**16.15:** (a) Increase power by taking more measurements. (b) If you increase $\alpha$, you make it easier to reject $H_0$, and hence increase power. (c) A value of $\mu = 10.2$ is even further from the stated value of $\mu = 10.1$ under $H_0$, so power increases.

**16.17:** The table below summarizes power as $\sigma$ changes. As $\sigma$ decreases, power increases. More precise measurements increase the researcher's ability to recognize a false null hypothesis.

| $\sigma$ | 0.10 | 0.05 | 0.025 |
|---|---|---|---|
| Power | 0.232 | 0.688 | 0.998 |

**16.19:** (a) All statistical methods are based on probability samples. We must have a random sample in order to apply them.

**16.21:** (b) Inference from a voluntary response sample is never reasonable. Online web surveys are voluntary response surveys.

**16.23:** (a) There is no control group. Any observed improvement may be due to the treatment, or may be due to another cause.

**16.25:** (a) The significance level ($\alpha$) is the probability of rejecting $H_0$ when $H_0$ is true.

**16.27:** (c) Power describes the test's ability to reject a false $H_0$.

**16.29:** We need to know that the samples taken from both populations (hunter-gatherers, agricultural) are random. Are the samples large? Recall that if the samples are very large, then even a small, practically insignificant difference in prevalence of color blindness in the two samples will be deemed statistically significant.

**16.31:** Many people might be reluctant to relate details of their sex lives, or perhaps some will be inclined to exaggerate. It would not be surprising that such an estimate would be biased, but this author cannot guess the direction of bias (will the estimate be too high or too low on average?).

**16.33:** The effect is greater if the sample is small. With a larger sample, the impact of any one value is small. That is, individual values are more influential in smaller samples.

**16.35:** Opinion—even expert opinion—unsupported by data is the weakest type of evidence, so the third description is level C. The second description refers to experiments (clinical trials) and large samples; that is the strongest evidence (level A). The first description is level B: stronger than opinion, but not as strong as experiments with large numbers of subjects.

**16.37:** (a) The P-value decreases (the evidence against $H_0$ becomes stronger). (b) The power increases (the test becomes better at distinguishing between the null and alternative hypotheses).

**16.39:** (a) The sample mean is $\bar{x} = 7.524$. The test statistic is $z = \dfrac{7.524 - 6}{2/\sqrt{5}} = 1.704$. The P-value is $P = 2P(Z \geq 1.704) = 0.0884$ (using software). This is not significant at the 5% level of significance, since the P-value is greater than the level of significance. (b) We would not reject 6 as a plausible value of $\mu$, even though (unknown to the researcher) $\mu = 5$. This isn't surprising, since $\bar{x} = 7.524$. In fact, we would not reject $H_0: \mu = 7$, either (and the P-value would be larger, suggesting less evidence against $H_0$).

**16.41:** (a) "Statistically insignificant" means that the differences observed were no more than might have been expected to occur by chance even if SES had no effect on LSAT results. (b) If the results are based on a small sample, then even if the null hypothesis were not true, the test might not be sensitive enough to detect the effect. Knowing the effects were small tells us that the test was not insignificant merely because of a small sample size.

**16.43:** We desire to know the sample size needed to estimate the mean to within margin of error 600 with 95% confidence. We know that $\sigma = 3000$ and $z^* = 1.96$, so the sample size required is $n = \left(\dfrac{z^*\sigma}{m}\right)^2 = \left(\dfrac{(1.96)(3000)}{600}\right)^2 = 96.04$. Remember always to round up. Hence, take $n = 97$.

**16.45:** (a) This test has a 20% chance of rejecting $H_0$ when the alternative is true. (b) If the test has 20% power, then when the alternative is true, it will fail to reject $H_0$ 80% of the time. (c) The sample sizes are very small, which typically leads to low-power tests.

**16.47:** From the applet, against the alternative $\mu = 8$, power = 0.609. Against the alternative $\mu = 10$, power = 0.994.

**16.49:** (a) Because the alternative is $\mu \neq 10.1$, we reject $H_0$ at the 5% level when $z \geq 1.96$ or $z \leq -1.96$. (b) Here, $z = \dfrac{\bar{x} - 10.1}{0.1/\sqrt{6}} = 24.4949\,(\bar{x} - 10.1)$. Hence, we reject $H_0$ if $24.4949(\bar{x} - 10.1) \leq -1.96$ or if $24.4949(\bar{x} - 10.1) \geq 1.96$. Equivalently (solving for $\bar{x}$), we reject $H_0$ if $\bar{x} \leq 10.02$ or $\bar{x} \geq 10.18$. (c) When $\mu = 10.15$, power represents the probability of rejecting $H_0$ when $\mu = 10.15$. We reject $H_0$ if $\bar{x} \leq 10.02$ or $\bar{x} \geq 10.18$, so power here is given by Power =

$$P(\bar{x} \leq 10.02) + P(\bar{x} \geq 10.18) = P\left(Z \leq \dfrac{10.02 - 10.15}{0.1/\sqrt{6}}\right) + P\left(Z \geq \dfrac{10.18 - 10.15}{0.1/\sqrt{6}}\right)$$

$= P(Z \leq -3.18) + P(Z \geq 0.74) = 0.0007 + 0.2297 = 0.2304$.

**16.51:** Power $= 1 - P(\text{Type II error}) = 1 - 0.14 = 0.86$.

**16.53:** (a) In the long run, this probability should be 0.05. Out of 100 simulated tests, the number of false rejections will have a binomial distribution with $n = 100$ and $p = 0.05$. Most students will see between 0 and 10 rejections. (b) If the power is 0.808, the probability of a Type II error is 0.192, so in the long run, this probability should be 0.192. Out of 100 simulated tests, the number of false non-rejections will have a binomial distribution with $n = 100$ and $p = 0.192$. Most students will see between 10 and 29 non-rejections.

# Chapter 17: From Exploration to Inference: Part II Review

Test Yourself Exercise Answers are answers or sketches. All of these problems are similar to ones found in Chapters 8–16, for which the solutions in this manual provide more detail.

17.1: (c) Hives with bees; Hives with no bees; No hives

17.3: (b)

17.5: (a) The subjects were not assigned to exercise type.

17.7: Many answers are possible. One possible lurking variable is "student attitude about purpose of college" (students with a view that college is about partying, rather than studying may be more likely to binge drink and more likely to have lower grades). Remember that a correct example of a lurking variable *must* be a variable that simultaneously drives both "GPA" and "binge drinking" together.

17.9: No doubt, Question A had 60% favoring a tax cut, while Question B had 22% favoring a tax cut. Question wording in A packages all government spending into an impersonal block that people don't relate to, or even strongly oppose. Question wording in B describes government spending with greater detail, and mentions government spending priorities most people care about (education, defense, etc.).

17.11: (b) the different types of movies.

17.13: (a) a matched–pairs experiment.

17.15: (d) probably biased. It's not a random sample, and those walking at night probably have a different view of campus safety than those than the campus community broadly defined.

17.17: (a) $1 - 0.66 - 0.21 - 0.07 - 0.04 = 0.02$.

17.19: $Y > 1$, or $Y \geq 2$. $P(Y \geq 2) = 1 - 0.26 = 0.74$.

17.21: (d) $1 - 0.33 = 0.67$.

17.23: $P(X \leq 2)$ is the probability of women giving birth to 2 or fewer children during their childbearing years. $P(X \leq 2) = 0.193 + 0.174 + 0.344 = 0.711$. 71.1% of women give birth to 2 or fewer children during their childbearing years.

17.25: $P(X \geq 3) = 0.181 + 0.074 + 0.034 = 0.289$.

17.27: (a) The height of the density curve is $1/5 = 0.2$, since the area under the density function must be 1.

17.29: (b) This is a personal probability.

17.31: (c) mean = 100, standard deviation = $15/\sqrt{60} = 1.94$ (rounded).

17.33: The answer in 17.30 would change, since this refers to the population distribution, which is now non-Normal. The answer in 17.31 would not change — the mean of $\bar{x}$ is 100, and the standard deviation of $\bar{x}$ is 1.94, regardless of the population distribution. The answer in 17.32 would, essentially, not change. The central limit theorem tells us that the sampling distribution of $\bar{x}$ is approximately Normal when $n$ is large enough (and 60 should be large enough), no matter what the population distribution.

17.37: (a) $11{,}479/14{,}099 = 0.8142$.  (b) $6457/(6457+1818) = 0.7803$.

17.39: (a) $1 - P(\text{failure}) = 1 - P(\text{both components fail}) = 1 - (0.20)(0.03) = 0.994$.

17.41: (c) This is a binomial distribution with $n = 1000$ trials and probability of success $p = 0.63$. Hence, the mean is $np = 1000(0.63) = 630$, and the standard deviation is $\sqrt{np(1-p)} = \sqrt{1000(0.63)(1-0.63)} = 15.27$.

17.43: (c) $357 \pm 1.96 \dfrac{50}{\sqrt{8}} = 322.35$ to $391.65$.

17.45: $357 \pm 1.282 \dfrac{50}{\sqrt{8}}$  $334.37$ to $379.63$.

17.47: (b) $172 \pm 18.03$ mg/dl.

17.49: (c) $n \geq \dfrac{(1.645)^2 (41)^2}{5^2} = 181.95$, which rounds to 182.

17.51: (c) $H_0 : \mu = 5$, $H_a : \mu \neq 5$.

17.53: (c) $\alpha = 0.10$ but not at $\alpha = 0.05$. The $P$-value is 0.0721.

17.55: (d) No more than 0.01. In Exercise 17.54, $P=0.0018$.

17.57: We test $H_0 : \mu = 100$ vs. $H_a : \mu < 100$; $z = \dfrac{87.6 - 100}{15/\sqrt{113}} = -8.79$; $P$-value $\approx 0$. Overwhelming evidence that the mean IQ for the very-low-birth-weight population is less than 100.

17.59: (c) The statement "no differences were seen" means that the observed differences were not statistically significant at the significance level used by the researchers.

# Solutions

89

**17.61:** Here, $r^2 = 0.61$ means that 61% of the total variability in number of wildfires is explained by our model (by knowing the year). If there is really no relationship between number of fires and year (a surrogate for population here), then an observed linear relationship in our data as strong as that observed ($r^2 = 0.61$) would have been very unlikely to occur by chance alone. It seems reasonable to conclude that "year" and "wildfires" are positively associated — fires have increased over time, suggesting that population (or changing weather) increases wildfires. However, a cause-and-effect conclusion is not possible.

## Supplementary Exercises

**17.63:** Placebos do work with real pain, so the placebo response tells nothing about physical basis of the pain. In fact, placebos work poorly in hypochondriacs. The survey is described in the April 3, 1979, edition of the *New York Times*.

**17.65:** (a) Increase. (b) Decrease. (c) Increase. (d) Decrease.

**Note:** *The first and third statements make an argument in favor of a national health insurance system, while the second and fourth suggest reasons to oppose it.*

**17.67:** (a) The factors are storage method (three levels: fresh, room temperature for one month, refrigerated for one month) and preparation method (two levels: cooked immediately, or after one hour). There are therefore six treatments (summarized in the table). The response variables are the tasters' color and flavor ratings. (b) Randomly allocate $n$ potatoes to each of the six groups, then compare ratings. (c) For each taster, randomly choose the order in which the fries are tasted.

|  | Cooked immediately | Wait one hour |
|---|---|---|
| Fresh | 1 | 2 |
| Stored | 3 | 4 |
| Refrigerated | 5 | 6 |

**17.69:** (a) All probabilities are between 0 and 1, and their sum is 1. (b) Let $R_1$ be Taster 1's rating and $R_2$ be Taster 2's rating. Add the probabilities on the diagonal (upper left to lower right): $P(R_1 = R_2) = 0.03 + 0.08 + 0.25 + 0.20 + 0.06 = 0.62$. (c) $P(R_1 > R_2) = 0.19$. This is the sum of the ten numbers in the "lower left" part of the table; the bottom four numbers from the first column, the bottom three from the second column, the bottom two from the third column, and the last number in the fourth column. These entries correspond to, for example, "Taster 2 gives a rating of 1, and Taster 1 gives a rating more than 1." $P(R_2 > R_1) = 0.19$; this is the sum of the ten numbers in the "upper right" part of the table. We could also find this by noting that this probability and the other two in this exercise must add to 1 (because they account for all of the entries in the table). Alternatively, noting that the matrix is symmetric (relative to the main diagonal), we must have $P(R_1 > R_2) = P(R_2 > R_1)$.

**17.71:** (a) Out of 100 BMIs, nearly all should be in the range $\mu \pm 3\sigma = 27 \pm 22.5 = 4.5$ to $49.5$. (b) The sample mean $\bar{x}$ has a $N(\mu, \sigma/\sqrt{100}) = N(27, 0.75)$ distribution, so nearly all such means should be in the range $27 \pm 3(0.75) = 27 \pm 2.25$, or $24.75$ to $29.25$.

**17.73:** (a) This is an observational study: Behavior is observed, but no treatment is imposed. (b) "Significant" means unlikely to happen by chance. In this study, researchers determined that the fact that light-to-moderate drinkers had a lower death rate than the other groups is evidence of a real difference, rather than mere coincidence. (c) For example, some nondrinkers might avoid drinking because of other health concerns.

**17.75:** (a) The stemplot confirms the description given in the text. (Arguably, there are two "mild outliers" visible in the stemplot, although the $1.5 \times IQR$ criterion only flags the highest as an outlier.) (b) STATE: Is there evidence that the mean body temperature for all healthy adults is not equal to $98.6°$? PLAN: Let $\mu$ be the mean body temperature. We test $H_0: \mu = 98.6°$ vs. $H_a: \mu \neq 98.6°$; the alternative is two-sided because we had no suspicion (before looking at the data) that $\mu$ might be higher or lower than $98.6°$. SOLVE: Assume we have a Normal distribution and an SRS. The average body temperature in our sample is $\bar{x} = 98.203°$, so the test statistic is $z = \frac{98.203 - 98.6}{0.7/\sqrt{20}} = -2.54$. The two-sided $P$-value is $P = 2P(Z < -2.54) = 0.0110$. CONCLUDE: We have fairly strong evidence—significant at $\alpha = 0.05$, but not at $\alpha = 0.01$— that mean body temperature is not equal to $98.6°$. (Specifically, the data suggests that mean body temperature is lower.)

```
96 | 8
97 | 344
97 | 888889
98 | 0133
98 | 5789
99 |
99 | 6
100| 2
```

**17.77:** STATE: What is the mean body temperature $\mu$ for healthy adults? PLAN: Let $\mu$ denote the mean body temperature for healthy adults. We will estimate $\mu$ by giving a 90% confidence interval. SOLVE: Assume we have a Normal distribution and an SRS. With $\bar{x} = 98.203$, $\sigma = 0.7$ and $z^* = 1.645$, our 90% confidence interval for $\mu$ is

$$98.203 \pm 1.645 \left(\frac{0.7}{\sqrt{20}}\right) = 98.203 \pm 0.257, \text{ or } 97.95° \text{ to } 98.46°.$$

CONCLUDE: We are 90% confident that the mean body temperature for healthy adults is between $97.95°$ and $98.46°$.

**17.79:** A low-power test has a small probability of rejecting the null hypothesis, at least for some alternatives. That is, we run a fairly high risk of making a Type II error (failing to reject $H_0$ when it is false) for such alternatives. Knowing that this can happen, we should not conclude that $H_0$ is "true" simply because we failed to reject it.

# Chapter 18: Inference about a Population Mean

18.1: The standard error of the mean is $s/\sqrt{n} = 63.9/\sqrt{1000} = 2.0207$ minutes.

18.3: (a) $t^* = 2.132$. (b) $t^* = 2.479$.

18.5: (a) df = 12 – 1 = 11, so $t^* = 2.201$. (b) df = 18 – 1 = 17, so $t^* = 2.898$.
(c) df = 6 – 1 = 5, so $t^* = 2.015$.

18.7: STATE: What is the mean percent $\mu$ of nitrogen in ancient air? PLAN: We will estimate $\mu$ with a 90% confidence interval. SOLVE: We are told to view the observations as an SRS. A stemplot shows some left-skew; however, for such a small sample, the data are not unreasonably skewed. There are no outliers. With $\bar{x} = 59.5889\%$ and $s = 6.2553\%$ nitrogen, we have df = 9–1 = 8, so for 90% confidence $t^* = 1.860$. The 90% confidence interval for $\mu$ is $59.5889 \pm 1.860 \dfrac{6.2553}{\sqrt{9}} = 59.5889 \pm 3.8783 = 55.71\%$ to $63.47\%$. CONCLUDE: We are 90% confident that the mean percent of nitrogen in ancient air is between 55.71% and 63.47%.

```
4 | 9
5 | 1
5 |
5 | 4
5 |
5 |
6 | 0
6 | 33
6 | 445
```

18.9: (a) df = 15 – 1 = 14. (b) $t = 2.12$ is bracketed by $t^* = 1.761$ (with two-tail probability 0.10) and $t^* = 2.145$ (with two-tail probability 0.05). Hence, since this is a two-sided significance test, $0.05 < P < 0.10$. (c) This test is significant at the 10% level since the $P < 0.10$. It is not significant at the 5% level since the $P > 0.05$. (d) From software, $P = 0.0524$.

18.11: This is a matched pairs design, since each chimp faced the problem in two variations. PLAN: Take $\mu$ to be the mean difference (collaboration required minus not). We test $H_0: \mu = 0$ vs. $H_a: \mu > 0$, using a one-sided alternative because the researchers that when collaboration is required, chimpanzees collaborate more frequently. SOLVE: We must assume that the monkeys can be regarded as an SRS. For each monkey, we compute the call minus pure tone differences; a stemplot of these differences (provided) shows no outliers or deviations from Normality. The mean and standard deviation are $\bar{x} = 70.378$ and $s = 88.447$ spikes/second, so $t = \dfrac{70.378 - 0}{88.447/\sqrt{37}} =$

4.84 with df = 36. This has a very small P-value: P < 0.0001. CONCLUDE: We have very strong evidence that the mean frequency of collaborations is greater when collaboration is required to solve this task.

```
-1 | 10
-0 | 8
-0 | 110
 0 | 011123333444
 0 | 56667
 1 | 0001244
 1 | 6677
 2 | 34
 2 | 6
```

**18.13:** The provided stemplot suggests that the distribution of nitrogen contents is heavily skewed. Although $t$ procedures are robust, they should not be used if the population being sampled from is this heavily skewed. In this case, $t$ procedures are not reliable.

```
0 | 00000000000111
0 | 2222233
0 | 44
0 |
0 |
1 |
1 |
1 | 4
```

**18.15:** (b) We virtually never know the value of $\sigma$. 18.15: (b) We virtually never know the value of $\sigma$.

**18.17:** (c) df = 25 – 1 = 24.

**18.19:** (a) 2.718. Here, df = 11.

**18.21:** (c) 72.7 to 97.3. The interval is computed as $85 \pm 3.250 \frac{12}{\sqrt{10}}$.

**18.23:** (b) If you sample 64 unmarried male students, and then sample 64 unmarried female students, no matching is present.

**18.25:** For the student group: $t = \frac{0.08 - 0}{0.37/\sqrt{12}} = 0.749$ (not 0.49, as stated). For the non-student group: $t = \frac{0.35 - 0}{0.37/\sqrt{12}} = 3.277$ (rather than 3.25, a difference that might be due to rounding error). From Table C, the first P-value is between 0.4 and 0.5 (software gives 0.47), and the second P-value is between 0.005 and 0.01 (software gives 0.007).

18.27: (a) The sample size is very large, so the only potential hazard is extreme skew. Since scores range only from 0 to 500, there is a limit to how skewed the distribution could be. (b) From Table C, we take $t^* = 2.581$ (df = 1000), or using software take $t^* = 2.5775$. For either value of $t^*$, the 99% confidence interval is $250 \pm 2.581 = 247.4$ to $252.6$. (c) Because the 99% confidence interval for $\mu$ does not contain 243 and is entirely above 243, we would fail to reject $H_0 : \mu = 243$ against the one-sided alternative hypothesis $H_a : \mu < 243$ at the 1% significance level.

18.29: (a) A subject's responses to the two treatments would not be independent. (b) We have $t = \dfrac{-0.326 - 0}{0.181/\sqrt{6}} = -4.41$. With df = 5, $P = 0.0069$, significant evidence of a difference.

18.31: (a) A stemplot is provided below, and suggests the presence of outliers. The sample is small and the stemplot is skewed, so use of $t$ procedures is not appropriate. (b) We will compute two confidence intervals, as called for. In the first interval, using all 9 observations, we have df = 8 and $t^* = 1.860$. For the second interval, removing the two outliers (1.15 and 1.35), df = 6 and $t^* = 1.943$. The two 90% confidence intervals are:

$$0.549 \pm 1.860 \left( \frac{0.403}{\sqrt{9}} \right) = 0.299 \text{ to } 0.799 \text{ grams, and}$$

$$0.349 \pm 1.943 \left( \frac{0.053}{\sqrt{7}} \right) = 0.310 \text{ to } 0.388 \text{ grams.}$$

(c) The confidence interval computed without the two outliers is much narrower. Using fewer data values reduces degrees of freedom (yielding a larger value of $t^*$). Also, smaller sample sizes yield larger margins of error. However, both of these effects are offset by removing two values far from the others — $s$ reduces from 0.403 to 0.053 by removing them.

```
 2 | 5
 3 | 3 3 5 8
 4 | 0 0
 5 |
 6 |
 7 |
 8 |
 9 |
10 |
11 | 5
12 |
13 | 5
```

18.33: (a) The stemplot provided clearly shows the high outlier mentioned in the text. (b) Let $\mu$ be the mean difference (control minus experimental) in healing rates. We test $H_0: \mu = 0$ vs. $H_a: \mu > 0$. The alternative hypothesis says that the control limb heals faster; that is, the healing rate is greater for the control limb than for the experimental limb. With all 12 differences: $\bar{x} = 6.417$ and $s = 10.7065$, so $t = \dfrac{6.417 - 0}{10.7065/\sqrt{12}} = 2.08$. With df = 11, $P = 0.0311$ (using software). Omitting the outlier: $\bar{x} = 4.182$ and $s = 7.7565$, so $t = \dfrac{4.182 - 0}{7.7565/\sqrt{11}} = 1.79$. With df = 10, $P = 0.052$. Hence, with all 12 differences there is greater evidence that the mean healing time is greater for the control limb. When we omit the outlier, the evidence is weaker.

```
-1 | 3
-0 | 6
-0 |
 0 | 12
 0 | 5789
 1 | 012
 1 |
 2 |
 2 |
 3 | 1
```

18.35: (a) A histogram of the sample is provided. The sample has a significant outlier, and indicates skew. We might consider applying $t$ procedures to the sample after removing the most extreme observation (37,786). (b) If we remove the largest observation, the remaining sample is not heavily skewed and has no outliers. Now we test $H_0: \mu = 7000$ vs. $H_a: \mu \neq 7000$. With the outlier removed, $\bar{x} = 11{,}555.16$ and $s = 6{,}095.015$. Hence, $t = \dfrac{11555.16 - 7000}{6095.015/\sqrt{19}} = 3.258$. With df = 18 with software, $P = 0.0044$ (this is a two-sided test). There is overwhelming evidence that the mean number of words per day of men at this university differs from 7000.

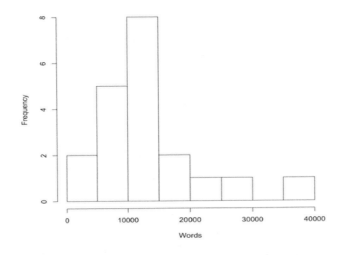

18.37: (a) The stemplot of differences shows an extreme right skew, and one or two high outliers. The $t$ procedures should not be used. (b) Some students might perform the test ($H_0: \mu = 0$ vs. $H_a: \mu > 0$) using $t$ procedures, despite the presence of strong skew and outliers in the sample. If so, they should find $\bar{x} = 156.36$, $s = 234.2952$, and $t = 2.213$, yielding $P = 0.0256$.

```
0 | 0 0 1 2 2 3 8
1 | 0
2 | 1
3 |
4 |
5 | 1
6 |
7 | 0
```

18.39: (a) We test $H_0: \mu = 0$ vs. $H_a: \mu > 0$, where $\mu$ is the mean difference (treated minus control). This is a one-sided test because the researchers have reason to believe that $CO_2$ will increase growth rate. (b) We have $\bar{x} = 1.916$ and $s = 1.050$, so $t = \dfrac{1.916 - 0}{1.050/\sqrt{3}} = 3.16$ with df = 2. Hence, $P = 0.044$. This is significant at the 5% significance level. (c) For very small samples, $t$ procedures should only be used when we can assume that the population is Normal. We have no way to assess the Normality of the population based on these four observations. Hence, the validity of the analysis in (b) is dubious.

18.41: The stemplot (not asked for) reveals that these data contain two extreme high outliers (5973 and 8015). Hence, $t$ procedures are not appropriate.

```
0 | 1123788
1 | 00115677899
2 | 01112458
3 |
4 |
5 | 9
6 |
7 |
8 | 0
```

18.43: (a) The mean and standard deviation are $\bar{x} = 48.25$ and $s = 40.24$ thousand barrels. From Table C, $t^* = 2.000$ (df = 60). Using software, with df = 63, $t^* = 1.998$. The 95% confidence interval for $\mu$ is $48.25 \pm 2.000 \left( \dfrac{40.24}{\sqrt{64}} \right) = 48.25 \pm 10.06 = 38.19$ to $58.31$ thousand barrels.

(Using the software version of $t^*$, the confidence interval is almost identical: 38.20 to 58.30 thousand barrels.) (b) The stemplot confirms the skewness and outliers described in the exercise. The two intervals have similar widths, but the new interval (using a computer-intensive method) is shifted higher by about 2000 barrels. Although $t$ procedures are fairly robust, we should be cautious about trusting the result in (a) because of the strong skew and outliers. The computer-intensive method may produce a more reliable interval.

```
0 | 00001111111111
0 | 2222222333333333333333
0 | 44444445555555
0 | 6666667
0 | 8899
1 | 01
1 |
1 | 5
1 |
1 | 9
2 | 0
```

18.45: PLAN: We will construct a 90% confidence interval for $\mu$, the mean percent of beetle-infected seeds. SOLVE: A stemplot shows a single-peaked and roughly symmetric distribution. We assume that the 28 plants can be viewed as an SRS of the population, so $t$ procedures are appropriate. We have $\bar{x} = 4.0786$ and $s = 2.0135$ percent. Using df = 27, the 90% confidence interval for $\mu$ is

$4.0786 \pm 1.703 \left( \dfrac{2.0135}{\sqrt{28}} \right) = 4.0786 \pm 0.648 = 3.43\%$ to $4.73\%$. CONCLUDE: The beetle infects less than 5% of seeds, so it is unlikely to be effective in controlling velvetleaf.

18.47: We have $\bar{x} = 0.5283$, $s = 0.4574$ and df = 5. Hence, a 95% confidence interval for the mean difference in T-cell counts after 20 days on blinatumomab is $0.5283 \pm 2.571 \left( \dfrac{0.4574}{\sqrt{6}} \right) =$ $0.5283 \pm 0.4801 = 0.0482$ to $1.0084$ thousand cells.

18.49: (a) For each subject, randomly select which knob (right or left) that subject should use first. (b) PLAN: We test $H_0 : \mu = 0$ vs. $H_a : \mu < 0$, where $\mu$ denotes the mean difference in time (right-thread time – left-thread time), so that $\mu < 0$ means "right-hand time is less than left-hand time on average". SOLVE: A stemplot of the differences gives no reason that $t$ procedures are not appropriate – the stemplot is roughly symmetric, with no outliers. We assume our sample can be viewed as an SRS. We have $\bar{x} = -13.32$ seconds and $s = 22.936$ seconds, so $t = \dfrac{-13.32 - 0}{22.936/\sqrt{25}} = -2.90$. With df = 24 we find $P = 0.0039$. CONCLUDE: We have very strong evidence (significant at the 1% level) that the mean difference really is negative — that is, the mean time for right-hand-thread knobs is less than the mean time for left-hand-thread knobs.

```
-5 | 2
-4 | 853
-3 | 511
-2 | 94
-1 | 66621
-0 | 74331
 0 | 02
 1 | 1
 2 | 03
 3 | 8
```

18.51: Refer to the solution in Exercise 18.49. With df = 24, $t^*$=1.711, so the confidence interval for $\mu$ is given by $-13.32 \pm 1.711\left(\dfrac{22.936}{\sqrt{25}}\right) = -13.32 \pm 7.85 = -21.2$ to $-5.5$ seconds. Now $\bar{x}_{RH}/\bar{x}_{LH} = 104.12/117.44 = 0.887$. Hence, right-handers working with right-handed knobs can accomplish the task in about 89% of the time needed by those working with left-handed knobs.

# Chapter 19: Two-Sample Problems

**19.1:** This is a matched–pairs design. Each couple is a matched pair.

**19.3:** This involves a single sample.

**19.5:** (a) If the loggers had known that a study would be done, they might have (consciously or subconsciously) cut down fewer trees than they typically would, in order to reduce the impact of logging. (b) State: Does logging significantly reduce the mean number of species in a plot after 8 years? PLAN: We test $H_0 : \mu_1 = \mu_2$ vs. $H_a : \mu_1 > \mu_2$, where $\mu_1$ is the mean number of species in unlogged plots and $\mu_2$ is the mean number of species in plots logged 8 years earlier. We use a one-sided alternative because we expect that logging reduces the number of tree species. SOLVE: We assume that the data come from SRSs of the two populations. Stemplots suggest some deviation from Normality, and a possible low outlier for the logged-plot counts, but there is not strong evidence of non-Normality in either sample. We proceed with the $t$ test for two samples. With $\bar{x}_1 = 17.50$, $\bar{x}_2 = 13.67$, $s_1 = 3.53$, $s_2 = 4.50$, $n_1 = 12$ and $n_2 = 9$: $SE = \sqrt{\dfrac{s_1^2}{n_1} + \dfrac{s_2^2}{n_2}} = 1.813$ and $t = \dfrac{\bar{x}_1 - \bar{x}_2}{SE} = 2.11$. Using df as the smaller of $9 - 1$ and $12 - 1$, we have df $= 8$, and $0.025 < P < 0.05$. Using software, df $= 14.8$ and $P = 0.026$. CONCLUDE: There is strong evidence that the mean number of species in unlogged plots is greater than that for logged plots 8 years after logging.

```
Unlogged   Logged
13 000      0 4
14          0
15 00       0
16          1 0
17          1 2
18 0        1 455
19 00       1 7
20 0        1 88
21 0
22 00
```

**19.7:** In Exercise 19.5, we saw that $\bar{x}_1 = 17.50$, $\bar{x}_2 = 13.67$, and $SE = 1.813$. Using df $= 8$, $t^* = 3.355$. Hence, a 99% confidence interval for the mean difference in number of species in unlogged and logged plots is $\bar{x}_1 - \bar{x}_2 \pm t^* SE = -2.253$ to $9.913$ species.

**19.9(Expanded):** (a) Back-to-back stemplots of the time data are shown below on left. They appear to be reasonably Normal, and the discussion in the exercise justifies our treating the data as independent SRSs, so we can use the $t$ procedures. We wish to test $H_0 : \mu_1 = \mu_2$ vs. $H_a : \mu_1 < \mu_2$, where $\mu_1$ is the is the population mean time in the restaurant with no scent, and $\mu_2$ is the mean time in the restaurant with a lavender odor. Here, With $\bar{x}_1 = 91.27$, $\bar{x}_2 = 105.700$, $s_1 = 14.930$, $s_2 = 13.105$, $n_1 = 30$ and $n_2 = 30$: $SE = \sqrt{\dfrac{s_1^2}{n_1} + \dfrac{s_2^2}{n_2}} = 3.627$ and $t = \dfrac{\bar{x}_1 - \bar{x}_2}{SE} = -3.98$. Using software, df $= 57.041$ and $P = 0.0001$. Using the more conservative df $= 29$ (lesser of 30–

1 and 30–1) and Table C, $P < 0.0005$. There is very strong evidence that customers spend more time on average in the restaurant when the lavender scent is present. (b) Back-to-back stemplots of the spending data are below on the right. The distributions are skewed and have many gaps. We wish to test $H_0 : \mu_1 = \mu_2$ vs. $H_a : \mu_1 < \mu_2$, where $\mu_1$ is the the population mean amount spent in the restaurant with no scent, and $\mu_2$ is the mean amount spent in the restaurant with lavender odor. Here, With $\bar{x}_1 = \$17.5133$, $\bar{x}_2 = \$21.1233$, $s_1 = \$2.3588$, $s_2 = \$2.3450$, $n_1 = 30$ and $n_2 = 30$: SE $= \sqrt{\dfrac{s_1^2}{n_1} + \dfrac{s_2^2}{n_2}} = \$0.6073$ and $t = \dfrac{\bar{x}_1 - \bar{x}_2}{\text{SE}} = -5.95$. Using software, df = 57.041 and $P < 0.0001$. Using the more conservative df = 29 and Table C, $P < 0.0005$. There is very strong evidence that customers spend more money on average when the lavender scent is present.

```
No scent     | Lavender         No scent       | Lavender
      98 | 6 |                            9 | 12 |
     322 | 7 |                              | 13 |
     965 | 7 | 6                            | 14 |
      44 | 8 |                 999999999999 | 15 |
    7765 | 8 | 89                           | 16 |
   32221 | 9 | 234                          | 17 |
      86 | 9 | 578              555555555555 | 18 | 5555555555
      31 |10 | 1234                         | 19 |
    9776 |10 | 5566788999                 5 | 20 | 7
         |11 | 4                          9 | 21 | 5599999999
      85 |11 | 6                            | 22 | 3558
       1 |12 | 14                           | 23 |
         |12 | 69                           | 24 | 99
         |13 |                            5 | 25 | 59
         |13 | 7
```

**19.11:** We have two small samples ($n_1 = n_2 = 4$), so the $t$ procedures are not reliable unless both distributions are Normal.

**19.13:** Here are the details of the calculations:

$$SE_F = \frac{12.6961}{\sqrt{31}} \doteq 2.2803$$

$$SE_M = \frac{12.2649}{\sqrt{47}} \doteq 1.7890$$

$$SE = \sqrt{SE_F^2 + SE_M^2} \doteq 2.8983$$

$$df = \frac{SE^4}{\dfrac{1}{30}\left(\dfrac{12.6961^2}{31}\right)^2 + \dfrac{1}{46}\left(\dfrac{12.2649^2}{47}\right)^2} = \frac{70.565}{1.1239} \doteq 62.8$$

$$t = \frac{55.5161 - 57.9149}{SE} \doteq -0.8276$$

**19.15:** Reading from the software output shown in the statement of Exercise 19.13, we find that there was no significant difference in mean Self-Concept Scale scores for men and women ($t = -0.8276$, df = 62.8, $P = 0.4110$).

**19.17:** (a) the two-sample $t$ test. We have two independent populations: females and males.

**19.19:** (b) confidence levels and $P$-values from the $t$ procedures are quite accurate even if the population distributions are not exactly Normal.

19.21: (b) $\dfrac{15.84 - 9.64}{\sqrt{\dfrac{3.43^2}{21} + \dfrac{8.65^2}{21}}} = 3.05$.

19.23: (a) We suspect that younger people use social networks more than older people, so this is a one-sided alternative.

19.25: (a) To test the belief that women talk more than men, we use a one-sided alternative. $H_0: \mu_M = \mu_F$ vs. $H_a: \mu_M < \mu_F$. (b)–(d) The small table below provides a summary of $t$ statistics, degrees of freedom, and $P$-values for both studies. The two sample $t$ statistic is computed as $t = \dfrac{\bar{x}_F - \bar{x}_M}{\sqrt{\dfrac{s_F^2}{n_F} + \dfrac{s_M^2}{n_M}}}$, and we the conservative approach for computing df as the smaller sample size, minus 1.

| Study | $t$ | df | Table C values | $P$-value |
|---|---|---|---|---|
| 1 | −0.248 | 55 | $\|t\| < 0.679$ | $P > 0.25$ |
| 2 | 1.507 | 19 | $1.328 < t < 1.729$ | $0.05 < P < 0.10$ |

Note that for Study 1 we reference df = 50 in Table C. (e) The first study gives no support to the belief that women talk more than men; the second study gives weak support, significant only at a relatively high significance level (say $\alpha = 0.10$).

19.27: (a) Call group 1 the Stress group, and group 2 the No stress group. Then, since SEM = $s/\sqrt{n}$, we have $s = \text{SEM}\sqrt{n}$. Hence, $s_1 = 3\sqrt{20} = 13.416$ and $s_2 = 2\sqrt{51} = 14.283$. (b) Using the conservative Option 2, df = 19 (the lesser of 20 and 51, minus 1). (c) We test $H_0: \mu_1 = \mu_2$ vs. $H_a: \mu_1 \neq \mu_2$. With $n_1 = 20$ and $n_2 = 51$, SE = $\sqrt{\dfrac{s_1^2}{n_1} + \dfrac{s_2^2}{n_2}} = 3.605$, and $t = \dfrac{\bar{x}_1 - \bar{x}_2}{\text{SE}} = \dfrac{26 - 32}{3.605} = -1.664$. With df = 19, using Table C, $0.10 < P < 0.20$. There is little evidence in support of a conclusion that mean weights of rats in stressful environments differ from those of rats without stress.

19.29: (a) A placebo is an inert pill that allows researchers to account for any psychological benefit (or detriment) the subject might get from taking a pill. (b) Neither the subjects nor the researchers who worked with them knew who was getting ginkgo extract; this prevents expectations or prejudices from affecting the evaluation of the effectiveness of the treatment. (c) SE = $\sqrt{\dfrac{0.01462^2}{21} + \dfrac{0.01549^2}{18}} = 0.0048$, so the two-sample test statistic is $t = \dfrac{0.06383 - 0.05342}{\text{SE}} = 2.147$. This is significant at the 5% level: $P = 0.0387$ (df = 35.35) or $0.04 < P < 0.05$ (df = 17). There is strong evidence that those who take gingko extract average more misses per line.

19.31: Let $\mu_1$ be mean for people with Asperger syndrome, and let and $\mu_2$ be the mean for people without Asperger syndrome. We test $H_0 : \mu_1 = \mu_2$ vs. $H_a : \mu_1 \neq \mu_2$. This is a one-sided test, as the researchers suspect that people with Asperger syndrome have different mean score than people without Asperger syndrome. Here $\bar{x}_1 = -0.001$, $\bar{x}_2 = 0.42$, $n_1 = 19$, and $n_2 = 17$. We are given SEM (standard error for the mean) values for each group. Since SEM = $s/\sqrt{n}$, we have $s = SEM\sqrt{n}$. Hence, $s_1 = 0.15\sqrt{19} = 0.6538$ and $s_2 = 0.17\sqrt{17} = 0.7009$. Hence, SE = $\sqrt{\frac{s_1^2}{n_1} + \frac{s_2^2}{n_2}} = 0.2267$ and $t = \frac{\bar{x}_1 - \bar{x}_2}{SE} = -1.857$. Using the conservative version for df (Option 2), df = 16 and $0.05 < P < 0.10$. Using software, df = 32.89 and $P = 0.0723$. There is strong evidence that the mean score for Asperger syndrome population is different than that of the non-Asperger population.

19.33: (a) The hypotheses are $H_0 : \mu_1 = \mu_2$ vs. $H_a : \mu_1 > \mu_2$, where $\mu_1$ is the mean gain among all coached students, and $\mu_2$ is the mean gain among uncoached students. We find SE = $\sqrt{\frac{59^2}{427} + \frac{52^2}{2733}} = 3.0235$ and $t = \frac{29 - 21}{3.0235} = 2.646$. Using the conservative approach, df = 426 is rounded down to df = 100 in Table C and we obtain $0.0025 < P < 0.005$. Using software, df = 534.45 and $P = 0.004$. There is evidence that coached students had a greater average increase. (b) The 99% confidence interval is $8 \pm t^*(3.0235)$ where $t^*$ equals 2.626 (using df = 100 with Table C) or 2.585 (df = 534.45 with software). This gives either 0.06 to 15.94 points, or 0.184 to 15.816 points, respectively. (c) Increasing one's score by 0 to 16 points is not likely to make a difference in being granted admission or scholarships from any colleges.

19.35: (a) Histograms for both data sets are provided below. Neither sample histogram suggests strong skew or presence of far outliers. Hence, $t$ procedures are reasonable here. (b) Let $\mu_1$ be the mean tip percentage when the forecast is good, and $\mu_2$ be the mean tip percentage when the forecast is bad. We have Here $\bar{x}_1 = 22.22$, $\bar{x}_2 = 18.19$, $s_1 = 1.955$, $s_2 = 2.105$, $n_1 = 20$, and $n_2 = 20$. We test $H_0 : \mu_1 = \mu_2$ vs. $H_a : \mu_1 \neq \mu_2$. Here, SE = $\sqrt{\frac{s_1^2}{n_1} + \frac{s_2^2}{n_2}} = 0.642$ and $t = \frac{\bar{x}_1 - \bar{x}_2}{SE} = 6.274$. Using df = 19 (the conservative Option 2) and Table C, we have $P < 0.001$. Using software, df = 37.8, and $P < 0.00001$. There is overwhelming evidence that the mean tip percentage differs between the two types of forecasts presented to patrons.

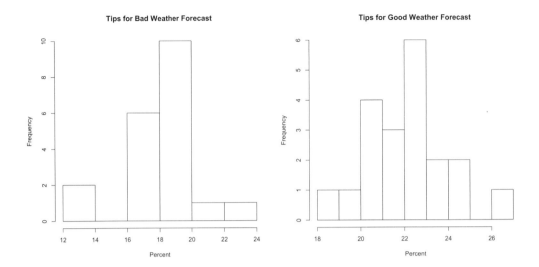

**19.37:** Refer to results in Exercise 19.35. Using df = 19, $t^* = 2.093$ and the 95% confidence interval for the difference in mean tip percentages between these two populations is 22.22−18.19 ± 2.093(0.642) = 4.03 ± 1.34 = 2.69 to 5.37 percent. Using df = 37.8 with software, the corresponding 95% confidence interval is 2.73% to 5.33%.

**19.39:** (a) The Hylite mean is greater than the Permafresh mean. (b) Shown are back-to-back stemplots for the two processes, which confirm that there are no extreme outliers. (c) We find SE = 1.334 and $t = -6.296$, for which the P-value is $0.002 < P < 0.005$ (using df = 4) or 0.0003 (using software, with df = 7.779). There is very strong evidence of a difference between the population means. As we might expect, the stronger process (Permafresh) is less resistant to wrinkles.

|  | n | $\bar{x}$ | s |
|---|---|---|---|
| Permafresh | 5 | 134.8 | 1.9235 |
| Hylite | 5 | 143.2 | 2.2804 |

| Permafresh |  | Hylite |
|---|---|---|
| 2 | 13 |  |
| 54 | 13 |  |
| 76 | 13 |  |
|  | 13 |  |
|  | 14 | 11 |
|  | 14 | 3 |
|  | 14 | 5 |
|  | 14 | 6 |

**19.41:** Summary statistics and background work is done in Exercise 19.39. The 90% confidence interval is $\bar{x}_1 - \bar{x}_2 \pm t^* SE$, where $t^* = 2.132$ (df = 4) or $t^* = 1.867$ (df = 7.779). This gives either

$$-8.4 \pm 2.844 = -11.244 \text{ to } -5.556 \text{ degrees (with df = 4), or}$$

$$-8.4 \pm 2.491 = -10.891 \text{ to } -5.909 \text{ degrees (with df = 7.779).}$$

**19.43:** This is a two-sample $t$ statistic, comparing two independent groups (supplemented and control). Using the conservative df = 5, $t = -1.05$ would have a P-value between 0.30 and 0.40, which (as the report said) is not significant. The test statistic $t = -1.05$ would not be significant for any value of df.

**19.45:** These are paired $t$ statistics: for each bird, the number of days behind the caterpillar peak was observed, and the $t$ values were computed based on the pairwise differences between the first and second years. For the control group, df = 5, and for the supplemented group, df = 6. The control $t$ is not significant (so the birds in that group did *not* "advance their laying date in the second year"), while the supplemented group $t$ is significant with one-sided $P = 0.0195$ (so those birds did change their laying date).

**19.47: PLAN:** We test $H_0 : \mu_1 = \mu_2$ vs. $H_a : \mu_1 > \mu_2$, where $\mu_1$ is the mean weight loss for adolescents in the gastric banding group, and $\mu_2$ is the mean time for the lifestyle intervention group. The alternative hypothesis is one-sided because the researcher suspects that gastric banding leads to greater average weight loss than lifestyle modification. **SOLVE:** We must assume that the data comes from an SRS of the intended population; we cannot check this with the data. The stemplots for each sample show no heavy skew and no outliers. With $\bar{x}_1 = 34.87$, $\bar{x}_2 = 3.01$, $s_1 = 18.12$, $s_2 = 13.22$, $n_1 = 24$ and $n_2 = 18$ (note that not all subjects completed the study). We find $SE = \sqrt{\frac{s_1^2}{n_1} + \frac{s_2^2}{n_2}} = 4.84$ and $t = \frac{314.0588 - 186.1176}{SE} = 6.59$, for which $P < 0.0005$ (df = 17) or $P < 0.00001$ (df = 39.98 using software). **CONCLUDE:** There is strong evidence that adolescents using gastric banding lose more weight on average than those that use lifestyle modification.

| Gastric Banding | Lifestyle Intervention |
|---|---|
| –0 \| 5 | –1 \| 762 |
| 0 \| | –0 \|44331 |
| 1 \| 359 | 0 \|12466 |
| 2 \| 02479 | 1 \|155 |
| 3 \| 12235679 | 2 \| 0 |
| 4 \| 139 | 3 \| 4 |
| 5 \| 37 | |
| 6 \| 4 | |
| 7 \| | |
| 8 \| 1 | |

**19.49: PLAN:** Compare mean length by testing $H_0: \mu_1 = \mu_2$ vs. $H_a: \mu_1 > \mu_2$, and by finding a 90% confidence interval for $\mu_1 - \mu_2$, where $\mu_1$ is the mean for the treatment population and $\mu_2$ is the mean for the control population. **SOLVE:** We must assume that we have two SRSs, and that the distributions of score improvements are Normal. The back-to-back stemplots of the differences ("after" minus "before") for the two groups; the samples are too small to assess Normality, but there are no outliers. With $\bar{x}_1 = 11.4$, $\bar{x}_2 = 8.25$, $s_1 = 3.1693$, $s_2 = 3.6936$, $n_1 = 10$ and $n_2 = 8$, we find SE $= \sqrt{\dfrac{3.1693^2}{10} + \dfrac{3.6936^2}{8}} = 1.646$ and $t = 1.914$. With either df = 7, $0.025 < P < 0.05$. With df =13.92 (software), $P = 0.0382$. The 90% confidence interval is $(11.4 - 8.25) \pm t^*$SE, where $t^* = 1.895$ (df = 7) or $t^* = 1.762$ (df = 13.92): either 0.03 to 6.27 points, or 0.25 to 6.05 points. **CONCLUDE:** We have fairly strong evidence that the encouraging subliminal message led to a greater improvement in math scores, on average. We are 90% confident that this increase is between 0.03 and 6.27 points (or 0.25 and 6.05 points).

| Treatment | | Control |
|---:|:---:|:---|
|   | 0 | 455 |
| 76 | 0 | 7 |
|   | 0 | 8 |
| 110 | 1 | 1 |
| 332 | 1 | 2 |
| 5 | 1 | 4 |
| 6 | 1 |   |

**19.51: PLAN:** Compare mean length by testing $H_0: \mu_1 = \mu_2$ vs. $H_a: \mu_1 \ne \mu_2$, and by finding a 95% confidence interval for $\mu_1 - \mu_2$, where $\mu_1$ is the mean for the Red population and $\mu_2$ is the mean for the Yellow population. **SOLVE:** We must assume that the data comes from an SRS. We also assume that the data are close to Normal. The back-to-back stemplots show some skewness in the red lengths, but the $t$ procedures should be reasonably safe. With $\bar{x}_1 = 39.7113$, $\bar{x}_2 = 36.1800$, $s_1 = 1.7988$, $s_2 = 0.9753$, $n_1 = 23$ and $n_2 = 15$, we find SE = 0.4518 and $t = 7.817$. With either df = 14 or df = 35.10, $P < 0.0001$. The 95% confidence interval is $(39.711 - 36.180) \pm t^*$SE, where $t^* = 2.145$ (df = 14) or $t^* = 2.030$ (df = 35.1): either 2.562 to 4.500 mm, or 2.614 to 4.448 mm. **CONCLUDE:** We have very strong evidence that the two varieties differ in mean length. We are 95% confident that the mean red length minus yellow length is between 2.562 and 4.500 mm (or 2.614 and 4.448 mm).

| Red | | Yellow |
|---:|:---:|:---|
|   | 34 | 56 |
|   | 35 | 146 |
|   | 36 | 0015678 |
| 9874 | 37 | 01 |
| 8722100 | 38 | 1 |
| 761 | 39 |   |
| 65 | 40 |   |
| 9964 | 41 |   |
| 10 | 42 |   |
| 0 | 43 |   |

# Chapter 20: Inference for a Population Proportion

20.1: (a) The population consists of all persons between the ages of 18 and 30 living in the United States. The parameter $p$ is the proportion of this population that prays at least once a week. Note here that the data are dichotomous – every individual in the population is a "yes" (prays at least once per week) or a "no" (does not pray at least once per week). This is the characteristic for problems in this chapter. (b) Our sample consists of 385 individuals, of which 287 answered "yes" (pray at least once per week). We estimate $p$ by $\hat{p} = \frac{247}{385} = 0.6416$. We estimate that about 64% of people in this population pray at least once per week.

20.3: (a) In the population, the proportion of young internet users using social media sites is given as $p = 0.70$. We envision taking a random sample of n = 1500 young Internet users and construct the sampling distribution of $\hat{p}$, the sample proportion using social media sites. With such a large sample size, the sampling distribution of $\hat{p}$ is approximately Normal with mean $p = 0.70$ and standard deviation $\sqrt{\frac{p(1-p)}{n}} = \sqrt{\frac{0.70(1-.070)}{1500}} = 0.0118$. (b) If $n = 6000$, the sampling distribution of $\hat{p}$ is approximately Normal with mean $p = 0.70$ and standard deviation $\sqrt{\frac{p(1-p)}{n}} = \sqrt{\frac{0.70(1-0.70)}{6000}} = 0.0059$. Notice that quadrupling the sample size (from 1500 to 6000) results in halving the standard deviation of $\hat{p}$ (0.0059 is one-half of 0.0118).

20.5: The survey excludes residents of the northern territories, as well as those who have no phones or have only cell phone service. (b) $\hat{p} = \frac{1288}{1505} = 0.8558$ so $SE_{\hat{p}} = \sqrt{\frac{\hat{p}(1-\hat{p})}{n}} = 0.009055$, and the 95% confidence interval is $0.8558 \pm 1.96(0.009055) = 0.8381$ to $0.8736$.

20.7: (a) Among the 14 observations, we have 11 successes and 3 failures. The number of successes and failures should both be at least 15 for the Normal approximation to be valid – large sample confidence intervals can't be used here. (b) However, the plus four confidence interval method can be used, since we have at least 10 observations. To construct the plus four estimate for $p$, we add 4 observations (2 successes and 2 failures), so we now have 18 observations: 13 successes and 5 failures. Now $\tilde{p} = \frac{11+2}{14+4} = \frac{13}{18} = 0.7222$. (c) Using the plus-four method, $SE_{\tilde{p}} = \sqrt{\frac{\tilde{p}(1-\tilde{p})}{n+4}} = \sqrt{\frac{0.7222(1-0.7222)}{18}} = 0.1056$. Hence, a 90% confidence interval for $p$ is $0.7222 \pm 1.645(0.1056) = 0.5485$ to $0.8959$. The confidence interval is quite wide (even with only 90% confidence level used) because the sample size is small.

20.9: (a) The sample proportion is $\hat{p} = \frac{20}{20} = 1$, so $SE_{\hat{p}} = \sqrt{\frac{\hat{p}(1-\hat{p})}{n}} = 0$. The margin of error would therefore be 0 (regardless of the confidence level), so large-sample methods give the useless interval 1 to 1. (b) The plus-four estimate is $\tilde{p} = \frac{20}{20} = 0.9167$, and $SE_{\tilde{p}} = \sqrt{\frac{\tilde{p}(1-\tilde{p})}{24}} = 0.0564$. A 95% confidence interval for $p$ is then $0.9167 \pm 1.96(0.0564) = 0.8062$ to $1.0272$. Since proportions can't exceed 1, we say that a 95% confidence interval for $p$ is 0.8061 to 1.

20.11: $n = \left(\frac{z^*}{m}\right)^2 p^*(1-p^*) = \left(\frac{1.645}{0.04}\right)^2 (0.75)(1-0.75) = 317.1$, so use $n = 318$.

20.13: STATE: We wonder if the proportion, $p$, of times the "best face" wins is more than 0.50. SOLVE: Let $p$ be the proportion of times the "best face" wins. We test $H_0: p = 0.50$ vs. $H_a: p > 0.50$, where the alternative hypothesis is one-sided because we wonder if the proportion has increased. Since the sample consists of 32 trials, we expect 16 "successes" (best face wins) and 16 "failures" (best face does not win). Hence, the sample is large enough to use the Normal approximation to describe the sampling distribution of $\hat{p}$. We assume the sample is an SRS. Here, $\hat{p} = \frac{22}{32} = 0.6875$ and $SE_{\hat{p}} = \sqrt{\frac{p_0(1-p_0)}{n}} = \sqrt{\frac{0.50(1-0.50)}{32}} = 0.0884$. Note the use of $p_0$ in the expression for $SE_{\hat{p}}$ above. With significance testing, we assume $H_0: p = 0.50$ is true, and this governs the sampling distribution of $\hat{p}$. Hence, $z = \frac{\hat{p} - p_0}{SE_{\hat{p}}} = \frac{0.6875 - 0.50}{0.0884} = 2.12$, and $P = P(z > 2.12) = 1 - 0.9830 = 0.0170$. CONCLUDE: There is strong evidence that the proportion of times the "best face" wins is more than 0.50.

20.15: (b) The sampling distribution of $\hat{p}$ has mean $p = 0.60$.

20.17: (c) $\hat{p} = \frac{1410}{3000} = 0.47$.

20.19: (c) $n = \left(\frac{z^*}{m}\right)^2 p^*(1-p^*) = \left(\frac{2.58}{0.02}\right)^2 (0.50)(1-0.50) = 4147.36$, round up to $n = 4148$.

20.21: (a) Sources of bias are not accounted for in a margin of error.

20.23: (c) $z = \frac{\hat{p} - p_0}{\sqrt{\frac{p_0(1-p_0)}{n}}} = \frac{0.53 - 0.50}{\sqrt{\frac{0.50(1-0.50)}{100}}} = 0.60$.

**20.25:** (a) The survey excludes those who have no phones or have only cell phone service. In fact, this is a growing portion of the population. Furthermore, and most troubling for polling organizations that conduct telephone surveys, younger people are more likely to have only a cell–phone. So, if younger people have different opinions on smoking than older people, the sample chosen only from people that have land lines is likely to be biased. (b) Note that we have plenty of successes and plenty of failures, so conditions for large-sample confidence interval are met. With the sample proportion feeling that smoking will probably shorten your life being $\hat{p} = \frac{848}{1010} = 0.8396$, the large sample 95% confidence interval is

$$\hat{p} \pm z^* \sqrt{\frac{\hat{p}(1-\hat{p})}{n}} = 0.8396 \pm 1.96 \sqrt{\frac{0.8396(1-0.8396)}{1010}} = 0.8170 \text{ to } 0.8622.$$

If we instead use the plus four method, $\tilde{p} = \frac{848 + 2}{1010 + 4} = 0.8383$, $SE_{\tilde{p}} = \sqrt{\frac{0.8383(1-0.8383)}{1010+4}} = 0.01156$, the margin of error is $1.96(0.01156) = 0.02266$, and the 95% confidence interval is 0.8156 to 0.8610. Notice that these confidence intervals agree closely because the sample is large.

**20.27:** (a) With $\hat{p} = \frac{848}{1010} = 0.8396$, $SE_{\hat{p}} = \sqrt{\frac{0.8396(1-0.8396)}{1010}} = 0.01155$, so the margin of error is $1.96 SE_{\hat{p}} = 0.02263 = 2.26\%$. (b) If instead $\hat{p} = 0.50$, then $SE_{\hat{p}} = \sqrt{\frac{0.50(1-0.50)}{1010}} = 0.01573$ and the margin of error for 95% confidence would be $1.96 SE_{\hat{p}} = 0.03084 = 3.08\%$. (c) For samples of about this size, the margin of error is no more than about $\pm 3\%$ no matter what $\hat{p}$ is. This is because the value of $SE_{\hat{p}}$ is largest when $\hat{p} = 0.50$. Note that many polling organizations do not report the maximum margin of error (about 3% in this case), but rather the estimated margin of error obtained using the observed value of $\hat{p}$ (about 2.3%, as computed in (a.)).

**20.29:** (a) The survey excludes residents of Alaska and Hawaii, and those who do not have cell–phone service. (b) We have 422 successes and 2063 failures, so the sample is large enough to use large-sample inference procedure. We have $\hat{p} = \frac{422}{2485} = 0.1698$, and $SE_{\hat{p}} = 0.0075$. For 90% confidence, the margin of error is $1.645 SE_{\hat{p}} = 0.0124$ and the confidence interval is 0.1574 to 0.1822, or 15.7% to 18.2%. Using the plus four method, $\tilde{p} = \frac{422 + 2}{2485 + 4} = 0.1703$, $SE_{\tilde{p}} = 0.0075$, the margin of error is $1.645 SE_{\tilde{p}} = 0.0124$, and the 90% confidence interval is 0.1579 to 0.1827, or 15.8% to 18.3%. (c) Perhaps people that use the cell–phone to search for information online are younger, and more sexually-related topics.

20.31: (a) In order to construct a large-sample confidence interval, we require at least 15 successes (swimming areas with unsafe levels of fecal coliform) and at least 15 failures (swimming areas with safe levels of fecal coliform). Here, we have 13 successes and 7 failures. In order to use the plus four confidence intervals, we require at least 90% confidence and at least 10 trials. Hence, conditions for using the plus four method are satisfied. (b) Now $\tilde{p} = \frac{13+2}{20+4} = 0.625$, and $SE_{\tilde{p}} = \sqrt{\frac{\tilde{p}(1-\tilde{p})}{24}} = 0.0988$. Hence, the margin of error for 90% confidence is $1.645(0.0988) = 0.1626$, and the 90% confidence interval for the proportion of swimming areas with unsafe coliform levels is 0.4624 to 0.7879, or 46.2% to 78.8%.

20.33: (a) Because the smallest number of total tax returns (i.e., the smallest population) is still more than 100 times the sample size, the margin of error will be (approximately) same for all states. (b) Yes, it will change—the sample taken from Wyoming will be about the same size, but the sample from, for example, California will be considerably larger, and therefore the margin of error will decrease.

20.35: (a) The margins of error are $1.96\sqrt{\frac{\hat{p}(1-\hat{p})}{100}} = 0.196\sqrt{\hat{p}(1-\hat{p})}$ (below). (b) With $n = 500$, the margins of error are $1.96\sqrt{\frac{\hat{p}(1-\hat{p})}{500}} = 0.088\sqrt{\hat{p}(1-\hat{p})}$. The new margins of error are less than half their former size.

| | p | 0.1 | 0.2 | 0.3 | 0.4 | 0.5 | 0.6 | 0.7 | 0.8 | 0.9 |
|---|---|---|---|---|---|---|---|---|---|---|
| (a) | m.e. | .0588 | .0784 | .0898 | .0960 | .0980 | .0960 | .0898 | .0784 | .0588 |
| (b) | m.e. | .0263 | .0351 | .0402 | .0429 | .0438 | .0429 | .0402 | .0351 | .0263 |

20.37: PLAN: We will give a 90% confidence interval for the proportion of all *Krameria cytisoides* shrubs that will resprout after fire. SOLVE: We assume that the 12 shrubs in the sample can be treated as an SRS. Because the number of resprouting shrubs is just 5, the conditions for a large sample interval are not met. Using the plus four method: $\tilde{p} = \frac{5+2}{12+4} = 0.4375$, $SE_{\tilde{p}} = 0.1240$, the margin of error is $1.645\, SE_{\tilde{p}} = 0.2040$, and the 90% confidence interval is 0.2335 to 0.6415. CONCLUDE: We are 90% confident that the proportion of *Krameria cytisoides* shrubs that will resprout after fire is between about 0.23 and 0.64.

20.39: PLAN: We will give a 95% confidence interval for $p$, the proportion of American adults who think that humans developed from earlier species of animals. SOLVE: We have an SRS with a very large sample size, so both large-sample and plus four methods can be used. We have $\hat{p} = \frac{594}{1484} = 0.4003$, $SE_{\hat{p}} = \sqrt{\frac{0.4003(1-0.4003)}{1484}} = 0.01272$, margin of error $1.96\, SE_{\hat{p}} = 0.02493$, and the 95% confidence interval is $0.4003 \pm 0.0249 = 0.3754$ to $0.4252$. Using the plus four method, we have $\tilde{p} = \frac{594+2}{1484+4} = 0.4005$, $SE_{\tilde{p}} = \sqrt{\frac{0.4005(1-0.4005)}{1484+4}} = 0.01270$, margin of error $1.96\text{SE}_{\tilde{p}} = 0.02489$. Hence the plus four 95% confidence interval is $0.4005 \pm 0.0249 =$ 0.3756 to 0.4254. CONCLUDE: We are 95% confident that the percent of American adults thinking that humans developed from earlier species of animals is between about 37.5% and 42.5%. Notice that the two confidence intervals computed agree closely because the sample is large.

20.41: PLAN: Let $p$ represent the proportion of American adults who think that humans developed from earlier species of animals. In order to assess the evidence for the claim that less than half of adults hold this belief, we will test $H_0: p = 0.50$ vs. $H_a: p < 0.50$. The test is one-sided because we wonder if the proportion is less than 0.50. SOLVE: We have an SRS with a very large sample size, so expected counts (successes and failures) are easily large enough to apply the large sample $z$ test. We have $\hat{p} = \dfrac{594}{1484} = 0.4003$. Because this is a significance test, we use the null-hypothesized value of $p$, $p_0 = 0.50$, in computing $\text{SE}_{\hat{p}} = \sqrt{\dfrac{0.50(1-0.50)}{1484}} = 0.01298$. Hence, the test statistic is $z = \dfrac{0.4003 - 0.50}{0.01298} = -7.68$, for which $P < 0.0001$.
CONCLUDE: We have overwhelming evidence that fewer than half of adults believe that humans developed from earlier species of animals. Random chance does not explain such a small sample proportion.

20.43: PLAN: We will give a 95% confidence interval for $p$, the proportion of Chick-fil-A orders correctly filled. SOLVE: We will assume that the 196 visits constitute a random sample of all possible visits. In our sample, we have 182 successes (correctly filled orders) and 14 failures (incorrectly filled orders). We will use the plus four method, since we do not have at least 15 failures: We have $\tilde{p} = \dfrac{182+2}{196+4} = 0.92$, $\text{SE}_{\tilde{p}} = 0.0192$, margin of error $1.96\text{SE}_{\tilde{p}} = 0.0376$, and the 95% confidence interval is 0.8824 to 0.9576. CONCLUDE: We are 95% confident that the proportion of orders filled correctly by Chick-fil-A is between 0.882 and 0.958, or 88.2% to 95.8%.

# Chapter 21: Comparing Two Proportions

**21.1:** PLAN: We construct a 95% confidence interval for $p_1 - p_2$, where $p_1$ denotes the proportion of younger people that text often, and $p_2$ denotes the proportion of older people that text often. SOLVE: We have two large samples: 625 younger people and 1917 older people. The number of successes in each sample (475 and 786, respectively) and the number of failures in each sample (150 and 1131) are large enough to use large-sample methods. We have $\hat{p}_1 = \frac{475}{625} = 0.76$, and $\hat{p}_2 = \frac{786}{1917} = 0.4100$. Now SE $= \sqrt{\frac{\hat{p}_1(1-\hat{p}_1)}{625} + \frac{\hat{p}_2(1-\hat{p}_2)}{1917}} = 0.0204$, so the margin of error for 95% confidence is 1.96(0.0204) = 0.0400 and a 95% confidence interval for the difference in proportions is (0.7600 – 0.4100) ± 0.0400 = 0.35 ± 0.04 = 0.31 to 0.39, or 31% to 39%. CONCLUDE: With 95% confidence, the proportion of younger people that text often exceeds that of older by somewhere between 0.31 and 0.39. Note that if you use the plus four interval, $\tilde{p}_1 = \frac{475+1}{625+2} = 0.7592$, $\tilde{p}_2 = \frac{786+1}{1917+2} = 0.4101$, SE $= \sqrt{\frac{\tilde{p}_1(1-\tilde{p}_1)}{625+2} + \frac{\tilde{p}_2(1-\tilde{p}_2)}{1917+2}} = 0.0204$, yielding (0.7592 – 0.4101) ± 1.96(0.0204) = 0.3091 to 0.3891. This interval is very close to the large-sample interval because the two sample sizes are large.

**21.3:** Let $p_1$ denote the proportion of males that meet recommended levels, and let $p_2$ denote the proportion for females. We have many successes and failures in both samples, so large sample methods are reasonable. Then $\hat{p}_1 = \frac{3594}{7881} = 0.4560$, and $\hat{p}_2 = \frac{2261}{8164} = 0.2769$. Then SE = 0.0075, and the margin of error is 2.576 SE = 0.0193. A 99% confidence interval for the difference in proportions between males and females meeting recommended levels of physical activity is 0.1598 to 0.1984, or 16.0% to 19.8%.

**21.5:** STATE: How much does microwaving crackers reduce checking? PLAN: Let $p_1$ denote the proportion of checking in the control group (no microwaving), and let $p_2$ denote the proportion of checking in the microwaved group. We give a 95% (plus four) confidence interval for $p_1 - p_2$. SOLVE: We cannot use the large sample method for computing a confidence interval here because we have no crackers with checking (successes) in the microwave group. To use plus four methods, we want samples of at least size 5; this condition is easily met here. The plus four estimates are $\tilde{p}_1 = \frac{16+1}{65+2} = 0.2537$ and $\tilde{p}_2 = \frac{0+1}{65+2} = 0.0149$. Hence, a plus four 95% confidence interval for $p_1 - p_2$ is $\tilde{p}_1 - \tilde{p}_2 \pm 1.96\sqrt{\frac{\tilde{p}_1(1-\tilde{p}_1)}{65+2} + \frac{\tilde{p}_2(1-\tilde{p}_2)}{65+2}} = 0.2388 \pm 0.1082$ = 0.1306 to 0.3470. CONCLUDE: We are 95% confident that microwaving reduces the proportion of crackers with checking by between about 13% and 35%.

**21.7: STATE:** Is helmet use less common among skiers and snowboarders with head injuries than skiers and snowboarders without head injuries? **PLAN:** Let $p_1$ be the proportion of injured alpine snowboarders or skiers that wear a helmet, and let $p_2$ be the proportion of uninjured skiers and snowboarders who wear helmets. We test $H_0: p_1 = p_2$ vs. $H_a: p_1 < p_2$. **SOLVE:** The smallest count is 96, so the significance testing procedure is safe. We find $\hat{p}_1 = \frac{96}{578} = 0.1661$ and $\hat{p}_2 = \frac{656}{2992} = 0.2193$. The pooled proportion is $\hat{p} = \frac{96+656}{578+2992} = 0.2106$. Then for the significance test, SE $= \sqrt{\hat{p}(1-\hat{p})\left(\frac{1}{578} + \frac{1}{2992}\right)} = 0.01853$. The test statistic is therefore $z = \frac{0.1661 - 0.2193}{\text{SE}} = -2.87$, and $P = 0.0021$. **CONCLUDE:** We have very strong evidence (significant at $\alpha = 0.01$) that skiers and snowboarders with head injuries are less likely to use helmets than skiers and snowboarders without head injuries. However, this is an observational study – people that do no wear helmets may ski or snowboard more hazardously than people that wear helmets – so the higher injury rate among non-helmet wearers may be due to a difference in behavior, rather than lack of protective helmet use.

**21.9:** (b) We look for evidence that the proportion for 2009 is lower than for 1999.

**21.11:** (b) $\hat{p} = \frac{511+592}{2411+2045} = 0.2475$, which rounds to 0.25.

**21.13:** (c) For a 95% confidence interval, the margin of error is

$$1.96\sqrt{\frac{\hat{p}_{1999}(1-\hat{p}_{1999})}{2411} + \frac{\hat{p}_{2009}(1-\hat{p}_{2009})}{2045}} = 0.026.$$

**21.15:** (b) We have only three failures in the treatment group, and only two successes in the control group.

**21.17:** (a) In this setting, either method is appropriate. We have 117 "successes" and 170 – 117 = 53 "failures" in the younger group. We have 152 "successes" 317 – 152 = 65 "failures" in the older group, so all counts are large enough. (b) Using the plus four method, as recommended, $\tilde{p}_1 = \frac{117+1}{170+2} = 0.6860$ and $\tilde{p}_2 = \frac{152+1}{317+2} = 0.4796$, and the 95% confidence interval is $\tilde{p}_1 - \tilde{p}_2 \pm 1.96\sqrt{\frac{\tilde{p}_1(1-\tilde{p}_1)}{172} + \frac{\tilde{p}_2(1-\tilde{p}_2)}{319}} = 0.2064 \pm 0.08841 = 0.1180$ to $0.2948$. Using the large-sample method, $\hat{p}_1 = \frac{117}{170} = 0.6882$, and $\hat{p}_2 = \frac{152}{317} = 0.4795$, and the 95% confidence interval is $\hat{p}_1 - \hat{p}_2 \pm 1.96\sqrt{\frac{\hat{p}_1(1-\hat{p}_1)}{170} + \frac{\hat{p}_2(1-\hat{p}_2)}{317}} = 0.2087 \pm 0.08873 = 0.1200$ to $0.2974$.

112    Chapter 21    Comparing Two Proportions

21.19: (a) Here, only the plus four method is appropriate. One of the counts is 0; for large-sample intervals, we want all counts to be at least 10, and for significance testing, we want all counts to be at least 5. (b) The sample size for the treatment group is 35, 24 of which have tumors; the sample size for the control group is 20, 1 of which has a tumor. (c) We have $\tilde{p}_1 = \frac{23+1}{33+2} = 0.6857$ and $\tilde{p}_2 = \frac{0+1}{18+2} = 0.05$. The plus four 99% confidence interval is

$$\tilde{p}_1 - \tilde{p}_2 \pm 2.576 \sqrt{\frac{\tilde{p}_1(1-\tilde{p}_1)}{33+2} + \frac{\tilde{p}_2(1-\tilde{p}_2)}{18+2}} = 0.6357 \pm 0.2380 = 0.3977 \text{ to } 0.8737.$$

We are 99% confidence that lowering DNA methylation increases the incidence of tumors by between about 40% and 87%.

21.21: (a) Let $p_1$ and $p_2$ be (respectively) the proportions of subjects in the music and no music groups that receive a passing grade on the Maryland HSA. We test $H_0 : p_1 = p_2$ vs. $H_a : p_1 \neq p_2$. For the music group $\hat{p}_1 = \frac{2818}{3239} = 0.870$. For the no music group, $\hat{p}_2 = \frac{2091}{2787} = 0.750$. The pooled estimate is $\hat{p} = \frac{2818 + 2091}{3239 + 2787} = 0.815$. Hence,

$$z = \frac{\hat{p}_1 - \hat{p}_2}{\sqrt{\hat{p}(1-\hat{p})\left(\frac{1}{3239} + \frac{1}{2787}\right)}} = 11.94.$$ An observed difference of $0.87 - 0.75 = 0.12$ in group proportions is much too large to be explained by chance alone, and $P < 0.001$. We have overwhelming evidence (or do we? See part (b).) that the proportion of music students passing the Maryland HSA is greater than that for the no music group. (b) and (c) This is an observational study — people that choose to (or can afford to) take music lessons differ in many ways from those that do not. Hence, we cannot conclude that music causes an improvement in Maryland HSA achievement.

21.23: The samples are so large, either confidence interval procedure is appropriate. Using the plus four method, we have $\tilde{p}_1 = \frac{2818+1}{3239+2} = 0.870$. and $\tilde{p}_2 = \frac{2091+1}{2787+2} = 0.750$. A 95% confidence interval for $p_1 - p_2$ is then $\tilde{p}_1 - \tilde{p}_2 \pm 1.96 \sqrt{\frac{\tilde{p}_1(1-\tilde{p}_1)}{3241} + \frac{\tilde{p}_2(1-\tilde{p}_2)}{2789}} = 0.100$ to $0.140$, or 10.0% to 14.0%. With such large samples, the large sample methods are appropriate also, but will yield virtually identical results: With $\hat{p}_1 = \frac{2818}{3239} = 0.870$. For the no music group, $\hat{p}_2 = \frac{2091}{2787} = 0.750$. $\hat{p}_1 - \hat{p}_2 \pm 1.96 \sqrt{\frac{\hat{p}_1(1-\hat{p}_1)}{3239} + \frac{\hat{p}_2(1-\hat{p}_2)}{2787}} = 0.100$ to $0.140$, or 10.0% to 14.0%.

21.25: (a) To test $H_0 : p_M = p_F$ vs. $H_a : p_M \neq p_F$, we find $\hat{p}_M = \dfrac{15}{106} = 0.1415$, $\hat{p}_F = \dfrac{7}{42} = 0.1667$, and $\hat{p} = 0.1486$. Then $\text{SE} = \sqrt{\hat{p}(1-\hat{p})\left(\dfrac{1}{106} + \dfrac{1}{42}\right)} = 0.06485$, so $z = \dfrac{\hat{p}_M - \hat{p}_F}{\text{SE}} = -0.39$. This gives $P = 0.6966$, which provides virtually no evidence of a difference in failure rates. (b) With larger sample sizes, we have $\hat{p}_M = \dfrac{450}{3180} = 0.1415$, $\hat{p}_F = \dfrac{210}{1260} = 0.1667$, and $\hat{p} = 0.1486$, but now $\text{SE} = \sqrt{\hat{p}(1-\hat{p})\left(\dfrac{1}{3180} + \dfrac{1}{1260}\right)} = 0.01184$, so $z = \dfrac{\hat{p}_M - \hat{p}_F}{\text{SE}} = -2.13$ and $P = 0.0332$.

Notice the important message: The pairs of sample proportions agree in both (a) and (b), but the sample sizes differ. The same absolute difference in proportions yields greater statistical significance (lower P-value) when samples are larger. (c) We are asked to construct two confidence intervals – one based on the smaller samples of part (a) and one based on the larger samples of part (b). In each case, we provide both large sample and plus four intervals, which are both appropriate here. First, for case (a), $\hat{p}_M = 0.1415$ and $\hat{p}_F = 0.1667$, so a 95% confidence interval for the difference is $\hat{p}_M - \hat{p}_F \pm 1.96\sqrt{\dfrac{\hat{p}_M(1-\hat{p}_M)}{106} + \dfrac{\hat{p}_F(1-\hat{p}_F)}{42}} = -0.1560$ to $0.1056$.

Using the plus four method, $\tilde{p}_M = \dfrac{15+1}{106+2} = 0.1481$ and $\tilde{p}_F = \dfrac{7+1}{42+2} = 0.1818$, so a 95% confidence interval for the difference is $\tilde{p}_M - \tilde{p}_F \pm 1.96\sqrt{\dfrac{\tilde{p}_M(1-\tilde{p}_M)}{108} + \dfrac{\tilde{p}_F(1-\tilde{p}_F)}{44}} = -0.0337 \pm 0.1322 = -0.1659$ to $0.0985$. For case (b), $\hat{p}_M = 0.1415$ and $\hat{p}_F = 0.1667$, but now $\tilde{p}_M = \dfrac{450+1}{3180+2} = 0.1417$ and $\tilde{p}_F = \dfrac{210+1}{1260+2} = 0.1672$. The resulting confidence intervals are then

$\hat{p}_M - \hat{p}_F \pm 1.96\sqrt{\dfrac{\hat{p}_M(1-\hat{p}_M)}{3180} + \dfrac{\hat{p}_F(1-\hat{p}_F)}{1260}} = -0.0491$ to $-0.0013$. The plus four interval is

$\tilde{p}_M - \tilde{p}_F \pm 1.96\sqrt{\dfrac{\tilde{p}_M(1-\tilde{p}_M)}{3182} + \dfrac{\tilde{p}_F(1-\tilde{p}_F)}{1262}} = -0.0494$ to $-0.0016$. Notice the important message here: Larger sample sizes reduce the margin of error, making it easier to detect a difference in population proportions. This was reflected as a lower P-value when doing significance testing, but more productively reflected as a smaller margin of error when constructing confidence intervals. The confidence interval allows one to asses practical significance, while the significance test allows one to assess only statistical significance.

**21.27: PLAN:** Let $p_1$ be the proportion of women who succeed, and let $p_2$ be that proportion of men who succeed. We test $H_0: p_1 = p_2$ vs. $H_a: p_1 \neq p_2$. **SOLVE:** The smallest count is 11, so the significance test should be safe. We find $\hat{p}_1 = \frac{23}{34} = 0.6765$ and $\hat{p}_2 = \frac{60}{89} = 0.6742$. The pooled proportion is $\hat{p} = \frac{23+60}{34+89} = 0.6748$, and SE $= \sqrt{\hat{p}(1-\hat{p})\left(\frac{1}{34} + \frac{1}{89}\right)} = 0.09445$. The test statistic is therefore $z = \frac{0.6765 - 0.6742}{\text{SE}} = 0.02$, for which $P = 0.9840$. **CONCLUDE:** We have no evidence to support a conclusion that women's and men's success rates differ.

**21.29: PLAN:** We construct a 99% confidence interval for $p_1 - p_2$, where $p_1$ denotes the proportion of people on Chantix who abstained from smoking, and $p_2$ is the corresponding proportion for the placebo population. **SOLVE:** The sample counts are 155 and 61 (successes for treatment and control, respectively) and 197 and 283 (failures for treatment and control, respectively), so the large-sample procedures are safe. Using the plus four procedure: $\tilde{p}_1 = \frac{155+1}{352+2} = 0.4407$, $\tilde{p}_2 = \frac{61+1}{344+2} = 0.1792$, and the 99% confidence interval is $\tilde{p}_1 - \tilde{p}_2 \pm 2.576\sqrt{\frac{\tilde{p}_1(1-\tilde{p}_1)}{354} + \frac{\tilde{p}_2(1-\tilde{p}_2)}{346}} = 0.2615 \pm 0.0863 = 0.1752$ to $0.3478$. Using the large-sample method, $\hat{p}_1 = \frac{155}{352} = 0.4403$, and $\hat{p}_2 = \frac{61}{344} = 0.1773$, and the 99% confidence interval is $\hat{p}_1 - \hat{p}_2 \pm 2.576\sqrt{\frac{\hat{p}_1(1-\hat{p}_1)}{352} + \frac{\hat{p}_2(1-\hat{p}_2)}{344}} = 0.2630 \pm 0.0864 = 0.1766$ to $0.3494$. Notice that the two methods produce very close intervals, since the two sample sizes are large.

**CONCLUDE:** We are 99% confident that the success rate for abstaining from smoking is between 17.5 and 34.8 percentage points higher for smokers using Chantix than for smokers on a placebo.

**21.31: PLAN:** Let $p_1$ be the proportion of people that will reject an unfair offer from another person, and $p_2$ be the proportion that would reject an unfair offer from a computer. We test $H_0: p_1 = p_2$ vs. $H_a: p_1 > p_2$. **SOLVE:** All counts are greater than 5, so the conditions for a significance test are met. The sample proportions are $\hat{p}_1 = \frac{18}{38} = 0.4737$ and $\hat{p}_2 = \frac{6}{38} = 0.1579$. The pooled proportion is $\hat{p} = \frac{18+6}{38+38} = 0.3158$, and SE $= \sqrt{\hat{p}(1-\hat{p})\left(\frac{1}{38} + \frac{1}{38}\right)} = 0.1066$. The test statistic is therefore $z = \frac{0.4737 - 0.1579}{\text{SE}} = 2.96$, for which $P = 0.0015$. **CONCLUDE:** There is very strong evidence that people are more likely to reject an unfair offer from another person than from a computer.

21.33: Let $p_1$ and $p_2$ be (respectively) the proportions of mice ready to breed in good acorn years and bad acorn years. We give a 90% confidence interval for $p_1 - p_2$. SOLVE: One count is only 7, and the guidelines for using the large-sample method call for all counts to be at least 10, so we use the plus four method. We have $\tilde{p}_1 = \dfrac{54+1}{72+2} = 0.7432$, and $\tilde{p}_2 = \dfrac{10+1}{17+2} = 0.5789$, so the plus four 90% confidence interval is $\tilde{p}_1 - \tilde{p}_2 \pm 1.645\sqrt{\dfrac{\tilde{p}_1(1-\tilde{p}_1)}{74} + \dfrac{\tilde{p}_2(1-\tilde{p}_2)}{19}} = 0.1643 \pm 0.2042 = -0.0399$ to $0.3685$. CONCLUDE: We are 90% confident that the proportion of mice ready to breed in good acorn years is between 0.04 lower than and 0.37 higher than the proportion in bad acorn years.

21.35: (a) This is an experiment because the researchers assigned subjects to the groups being compared. (b) PLAN: Let $p_1$ and $p_2$ be (respectively) the proportions that have an RV infection for the HL+ group and control group. We test $H_0 : p_1 = p_2$ vs. $H_a : p_1 < p_2$, because we wonder if the proportion with infection is lower in the HL+ group than in the control group. SOLVE: We have large enough counts to use large-sample significance testing procedure safely. Now $\hat{p}_1 = \dfrac{49}{49+67} = 0.4224$, $\hat{p}_2 = \dfrac{49}{49+47} = 0.5104$, and $\hat{p} = \dfrac{49+49}{116+96} = 0.4623$. Hence, $\text{SE} = \sqrt{\hat{p}(1-\hat{p})\left(\dfrac{1}{116} + \dfrac{1}{96}\right)} = 0.0688$. The test statistic is therefore $z = \dfrac{0.4224 - 0.5104}{\text{SE}} = -1.28$, for which $P = 0.1003$. CONCLUDE: There is some evidence, but not very much evidence against the null hypothesis. There is little evidence to conclude that the proportion of HL+ users with a rhinovirus infection is less than that for non-HL+ users.

# Chapter 22: Inference about Variables: Part III Review

Test Yourself Exercise Solutions contain only answers or sketches of answers. All of these problems are similar to ones found in Chapters 18–21, for which the solutions in this manual provide more detail.

22.1: (c) The margin of error is $2.056(9.3)/\sqrt{27} = 3.7$.

22.3: (b) $t = 2.023$, df = 13.

22.5: (d) Our estimate is $\hat{p} = 1926/7028 = 0.274$.

22.7: (d) The standard error is $SE = \sqrt{\dfrac{0.148(1-0.148)}{6889} + \dfrac{0.274(1-0.274)}{7028}} = 0.0068$.

22.9: (a) The standard error is 0.0124. (b) A 95% confidence interval is 0.336 to 0.384.

22.11: (b) The margin of error is $2.005(3.2)/\sqrt{55} = 0.865$.

22.13: (a) df is the lesser of $(55 - 1)$ and $(200 - 1)$.

22.15: With such large samples, $t$ procedures are reasonable.

22.17: (b) The margin of error is 3.52. The point estimate is $11.4 - 6.7 = 4.7$.

22.19: (c) Plus four confidence intervals are reliable for samples of 5 or more in each group.

2.21: (b) $\hat{p} = 225/757 = 0.297$.

22.23: (b) $0.297 \pm 1.645(0.017)$

22.25: (c) It seems reasonable that the researchers suspect that VLBW babies are less likely to graduate from high school.

22.27: (b) $z = \dfrac{0.7397 - 0.8283}{\sqrt{\hat{p}(1-\hat{p})\left(\dfrac{1}{242} + \dfrac{1}{233}\right)}} = -2.34$.

22.29: (d) $t = \dfrac{86.2 - 89.8}{\sqrt{\dfrac{13.4^2}{38} + \dfrac{14^2}{54}}} = -1.25$, and the test is two-sided.

22.31: (b) $z = \dfrac{0.379 - 0.41}{\sqrt{0.41(1-0.41)/348}} = -1.18$, so $P = 0.1190$.

# Solutions

22.33: $0.58 \pm 1.645\sqrt{\dfrac{0.58(1-0.58)}{634}} = 0.58 \pm 0.03 = 0.55$ to $0.61$. A plus four interval will agree to three decimal places and would also be appropriate.

22.35: We test $H_0: p_b = p_w$ vs. $H_a: p_b \neq p_w$, where $p_b$ denotes the proportion of young black people that believe rap music videos contain too many references to sex, and $p_w$ denotes the corresponding proportion for young white people. With $\hat{p} = [0.72(634)+0.68(567)]/(634+567) = 0.701$, $z = \dfrac{0.72-0.68}{\sqrt{0.70(1-0.70)\left(\dfrac{1}{634}+\dfrac{1}{567}\right)}} = 1.51$ and $P = 2P(Z>1.51) = 0.131$. Note the two-sided $P$-value here. There is little evidence of a difference between black and white young people in the proportion believing that rap music videos contain too many references to sex.

22.37: $t = \dfrac{193-174}{\sqrt{\dfrac{68^2}{26}+\dfrac{44^2}{23}}} = 1.174$, and df = 22, the lesser of $23-1=22$ and $26-1=25$.

22.39: We must assume that each sample is an SRS taken from its respective populations (clinic dogs and pet dogs). We must also assume that the populations (cholesterol levels of pet dogs and cholesterol levels of clinic dogs) are Normal.

22.41: We have a random sample of adults, and each adult rates several different behaviors. We compare the average rating for speeding to the average rating of noisy neighbors. Since each adult responds to both ratings, this is a matched pairs test.

22.43: If the sample can be viewed as an SRS, a $t$ confidence interval for a population mean.

22.45: We are comparing the male and female responses on a quantitative measurement (how much attractiveness of former partner mattered) for 40 couples. Matched pairs $t$ test or confidence interval.

22.47: The response rate for the survey was only about 20% (427/2100 = 0.203), which might make the conclusions unreliable.

22.49: Each of a monkey's six trials are not independent. If a monkey prefers silence, it will almost certainly spend more time in the silent arm of the cage each time it is tested.

22.51: (a) PLAN: We will find a 99% confidence interval for $p_2 - p_1$, where $p_1$ is the proportion of subjects on Gardasin that get cancer, and $p_2$ is the corresponding proportion for the control group. SOLVE: We assume that we have SRSs from each population. Because there were no cases of cervical cancer in the Gardasil group, we should use the plus four procedure. We have $\tilde{p}_1 = \dfrac{0+1}{8487+2} = 0.000118$, and $\tilde{p}_2 = \dfrac{32+1}{8460+2} = 0.003900$. A 99% confidence interval for $p_2 - p_1$ is then given by $\tilde{p}_2 - \tilde{p}_1 \pm 2.576\sqrt{\dfrac{\tilde{p}_1(1-\tilde{p}_1)}{8489} + \dfrac{\tilde{p}_2(1-\tilde{p}_2)}{8462}} = 0.003782 \pm 0.001772 = 0.0020$ to $0.0056$. (b) PLAN: Now let $p_1$ denote the proportion in the Gardasil group with genital warts, and let $p_2$ be the corresponding proportion for the control group. We find a 99% confidence interval for $p_2 - p_1$. SOLVE: Because we have fewer than 10 "successes" in the Gardasil group, conditions for using the large-sample interval are not met. However, we can use the plus four interval. We find that $\tilde{p}_1 = 0.000253$ and $\tilde{p}_2 = 0.011644$. A 99% confidence interval for $p_2 - p_1$ is then $0.0082$ to $0.0145$. (c) CONCLUDE: Gardasil is seen to be effective in reducing the risk of both cervical cancer (by between 0.0020 and 0.0056, with 99% confidence) and genital warts (by between 0.0082 and 0.0145, with 99% confidence).

22.53: PLAN: We test $H_0: \mu_1 = \mu_2$ vs. $H_a: \mu_1 < \mu_2$, where $\mu_1$ is the mean number of new leaves on plants from the control population, and $\mu_2$ is the mean for the nitrogen population. SOLVE: With $\bar{x}_1 = 13.2857$, $\bar{x}_2 = 15.6250$, $s_1 = 2.0587$, $s_2 = 1.6850$, $n_1 = 7$ and $n_2 = 8$, we find SE = $\sqrt{\dfrac{2.0587^2}{7} + \dfrac{1.6850^2}{8}} = 0.9800$ and $t = \dfrac{13.3857 - 15.6250}{SE} = -2.387$. With Option 2, df = 6 and $P = 0.0271$. Or, using Option 1, df = 11.66 and $P = 0.0175$. CONCLUDE: We have strong evidence that nitrogen increases the formation of new leaves.

22.55: PLAN: Do a two-sided test because we have no advance claim about the direction of the difference: $H_0: \mu_1 = \mu_2$ vs. $H_a: \mu_1 \neq \mu_2$. SOLVE: We view the data as coming from two SRSs; the distributions show no strong departures from Normality. The means and standard deviations of the lightness scores are: $\bar{x}_1 = 48.9513$ and $s_1 = 0.2154$ (cotton) and $\bar{x}_2 = 41.6488$ and $s_2 = 0.3922$ (ramie). Ramie is darker, having a lower score for lightness. We find SE = 0.1582 and $t = 46.16$. With either df = 7 or df = 10.87 (software), $P \approx 0$. CONCLUDE: There is overwhelming evidence that cotton is lighter than ramie.

22.57: PLAN: We test $H_0: \mu_1 = \mu_2$ vs. $H_a: \mu_1 \neq \mu_2$. SOLVE: We are told that the samples may be regarded as SRSs from their respective populations. Back-to-back stemplots show that $t$ procedures are reasonably safe, since both distributions are only slightly skewed, with no outliers, and with fairly large sample sizes. We have $\bar{x}_1 = 4.1769$, $s_1 = 2.0261$ and $n_1 = 65$ (parent allows drinking) and $\bar{x}_2 = 4.5517$ and $s_2 = 2.4251$ and $n_2 = 29$ (parent does not allow drinking). Hence, SE = 0.5157 and $t = \dfrac{\bar{x}_1 - \bar{x}_2}{SE} = -0.727$. This is very close to zero, so we will certainly not reject the null hypothesis. Indeed, with df = 46.19 (software), $P = 0.47$. CONCLUDE: There is no significant difference in the mean number of drinks between female students with a parent that allows drinking and those whose parents do not allow drinking.

| Parent allows drinking | | No parent allows drinking |
|---|---|---|
| 00000 | 1 | 000 |
| 55555550000 | 2 | 0 |
| 5500000000000 | 3 | 00000055 |
| 500000000000 | 4 | 000000 |
| 00000000 | 5 | 000 |
| 5000000 | 6 | 0 |
| 00000 | 7 | 000 |
| 00 | 8 | 0 |
| 0 | 9 | 00 |
| 0 | 10 | 0 |

22.59: (a) Stemplots are provided. The diabetic potentials appear to be larger. (b) PLAN: We test $H_0: \mu_1 = \mu_2$ vs. $H_a: \mu_1 \ne \mu_2$, where $\mu_1$ is the mean potential for diabetics, and $\mu_2$ is the mean for the normal population. SOLVE: We assume we have two SRSs; the distributions appear to be safe for $t$ procedures, since there does not appear to be heavy skew or any outliers. With $\bar{x}_1$ = 13.0896, $\bar{x}_2$ = 9.5222, $s_1$ = 4.8391, $s_2$ = 2.5765, $n_1$ = 24 and $n_2$ = 18, we find SE = 1.1595 and $t$ = 3.077. With Option 2, df = 17 and $0.005 < P < 0.01$. Or, using Option 1, df = 36.6, and $P$ = 0.0040. CONCLUDE: We have strong evidence that the electric potential in diabetic mice is greater than the potential in normal mice. (c) If we remove the outlier, the diabetic mouse statistics change: $\bar{x}_1$ = 13.6130, $s_1$ = 4.1959, $n_1$ = 23. Now SE = 1.065 and $t$ = 3.841. With df = 16, $0.001 < P < 0.002$. With df = 37.15, $P$ = 0.0005. CONCLUDE: With the outlier removed, the evidence that diabetic mice have higher mean electric potential is even stronger.

| Diabetic | | Normal |
|---|---|---|
| | 1 | 0 |
| | 0 | |
| | 0 | |
| | 0 | 4 |
| 7 | 0 | 6777 |
| 988 | 0 | 8888999 |
| 1000000 | 1 | 00 |
| 3 | 1 | 233 |
| 5444 | 1 | 4 |
| 76 | 1 | |
| 9988 | 1 | |
| | 2 | |
| 2 | 2 | |

22.61: PLAN: We give a 95% confidence interval for $\mu$, the mean date on which the tripod falls through the ice. SOLVE: We assume that the data can be viewed as an SRS of fall-through times, and that the distribution is roughly Normal. We find $n$ = 91, $\bar{x}$ = 15.3736 and $s$ = 5.9789 days. Using df = 90, a 95% confidence interval is given by 14.13 to 16.62 days. CONCLUDE: We are 95% confident that the mean number of days for the tripod to fall through the ice is 14.13 days to 16.62 days from April 20 or between May 3 and May 5.

22.63: (a) "SEM" stands for "standard error of the mean"; SEM = $s/\sqrt{n}$. (b) Two-sample $t$-tests were done, because there are two separate, independent groups of mice. (c) The observed differences between the two groups of mice were so large that it would be unlikely to occur by chance alone if the two groups were the same in average. Specifically, if the two population means were the same, if we repeated the experiment, an observed difference in sample means as large as this would occur by chance alone less than 0.5% of the time, which is less often than the 5% level of significance criterion. This is strong evidence that the two population means differ. We conclude that both the insulin and glucose levels in blood plasma of P2-/- mice and wild type mice differ.

# Chapter 23: Two Categorical Variables: The Chi-Square Test

23.1: (a) The table provided gives percentages in each category for Facebook use frequency, and for each campus. As an example, there are 978 surveyed students from the University Park campus. Of these, the proportion of University Park campus students that do not use Facebook is $68/978 = 0.0695$, which rounds to 0.070 and is represented as 7% in the table. Similarly, 17.9% of those surveyed at the Commonwealth campus use Facebook at least once a week. (b) A side-by-side bar graph reveals that students on the main campus are much more likely to use Facebook at least daily, while commonwealth campus students are more likely not to use it at all. Overall, it seems that University Park students tend to use Facebook more often than Commonwealth students.

|  | Univ. Park | Commonwealth |
|---|---|---|
| Do not use Facebook | 7.0% | 28.3% |
| Several times a month or less | 5.6% | 8.7% |
| At least once a week | 22.0% | 17.9% |
| At least once a day | 65.4% | 45.0% |

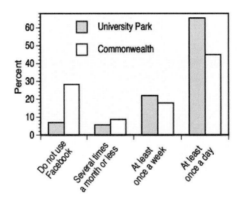

23.3: (a) In this problem, we compare the campuses on only the proportions that do not use Facebook. We test $H_0: p_1 = p_2$ vs. $H_a: p_1 \neq p_2$ for the proportions not using Facebook, we have $\hat{p}_1 = 0.0695$ (as illustrated in solution to Exercise 23.1) and $\hat{p}_2 = 0.2834$. The pooled proportion is $\hat{p} = \dfrac{68 + 248}{978 + 875} = 0.1705$, and the standard error is SE = 0.01750, so $z = -12.22$, for which $P$ is close to zero. This is overwhelming evidence against the null hypothesis. We conclude that the proportion of students not using Facebook differs between these two campus locations. (b) Here we test $H_0: p_1 = p_2$ vs. $H_a: p_1 \neq p_2$ for the proportions that use Facebook at least weekly. We have $\hat{p}_1 = 0.2198$ and $\hat{p}_2 = 0.1794$. The pooled proportion is $\hat{p} = 0.2008$ and the standard error is SE = 0.01864, so $z = 2.17$, for which $P = 0.0300$. There is strong evidence that the proportion of students using Facebook at least weekly differs between these two campus locations. (c) If we did four individual tests, we would not know how confident we could be in all four results when taken together.

23.5: (a) Expected counts are below observed counts in the table provided. For example, for students at the Commonwealth campus using Facebook Monthly, the expected count is $\frac{(131)(627)}{1537} = 53.44$. Similarly, for students at the University Park campus, the expected count using Facebook at least daily is $\frac{(910)(627)}{1537} = 612.19$. Notice that he expected counts add up to the same values as the observed counts by row, by column and overall. (b) We find that Commonwealth students actually use Facebook less than once weekly more often than we would expect (76 observed, 53.44 expected). Also, Commonwealth students use Facebook daily less often than we would expect (394 observed, 421.81 expected). In general, the pattern of lack of fit depicted in the table yields more students in the Commonwealth campus using Facebook less often than expected, and students in the University Park campus using Facebook more often than expected.

```
Minitab output
            UPark    Cwlth    Total
Monthly     55       76       131
            77.56    53.44

Weekly      215      157      372
            220.25   151.75

Daily       640      394      1034
            612.19   421.81

Total       910      627      1537
```

23.7: (a) All expected counts are well above 5 (the smallest is 53.44). (b) We test $H_0$: There is no relationship between setting and Facebook use vs. $H_a$: There is a relationship between campus and Facebook use. Using software, we have $\chi^2 = 19.489$ and $P < 0.0005$. (c) The largest contributions come from the first row, reflecting the fact that monthly use is lower among University Park students and higher among commonwealth students.

23.9: PLAN: We will carry out a chi-square test for association between education level and opinion about astrology. We test $H_0$: There is no relationship between education level and astrology opinion vs. $H_a$: There is some relationship between education level and astrology opinion. SOLVE: Examining the output provided in Figure 23.5, we see that all expected cell counts are greater than 5 and all observed cell counts are at least 1, so conditions for use of the chi-square test are satisfied. We see that $\chi^2 = 7.244$ and $P = 0.027$. CONCLUDE: There is strong evidence of an association between education level and opinion of astrology. Examining the table, we note that for people with Graduate degrees, more than expected felt that astrology is not scientific, while fewer than expected believed that astrology is scientific. For people with a Junior College degree, more than expected believed that astrology is scientific, and fewer than expected felt that astrology is not scientific.

23.11: (a) df $= (r-1)(c-1) = (3-1)(2-1) = 2$. (b) The largest critical value shown for df $= 2$ is 15.20; since the computed value (19.489) is greater than this, we conclude that $P < 0.0005$. (c) With $r = 4$ and $c = 2$, the appropriate degrees of freedom would be df $= 3$.

**23.13:** We test $H_0: p_1 = p_2 = p_3 = \frac{1}{3}$ vs. $H_a$: Not all three are equally likely. There were 53 bird strikes in all, so the expected counts are each $53 \times \frac{1}{3} = 17.67$. The chi-square statistic is then

$$\chi^2 = \sum \frac{(\text{observed count} - 17.67)^2}{17.67} = \frac{(31-17.67)^2}{17.67} + \frac{(14-17.67)^2}{17.67} + \frac{(8-17.67)^2}{17.67} = 10.06 +$$

$0.76 + 5.29 = 16.11$. The degrees of freedom are df = 2. From Table D, $\chi^2 = 16.11$ falls beyond the 0.005 critical value, so $P < 0.005$. There is very strong evidence that the three tilts differ. The data and terms of the statistic show that more birds than expected strike the vertical window and fewer than expected strike the 40-degree window.

**23.15:** Letting $p_1$ denote the proportion of people aged 16 to 29 not wearing seatbelts, and similarly defining $p_2$ and $p_3$ for the other age groups, we test $H_0: p_1 = p_2 = p_3$, vs. $H_a$: Not all proportions are equal. The details of the computation are shown below. The expected counts are found by multiplying the expected frequencies by 803 (the total number of observations).

|  | Expected Frequency | Observed Count | Expected Count | $O - E$ | $\frac{(O-E)^2}{E}$ |
|---|---|---|---|---|---|
| 16 to 29 | 0.328 | 401 | 263.384 | 137.616 | 71.9032 |
| 30 to 59 | 0.594 | 382 | 476.982 | −94.982 | 18.9139 |
| 60 or older | 0.078 | 20 | 62.634 | −42.634 | 29.0203 |
|  |  | 803 |  |  | 119.8374 |

The difference is significant: $\chi^2 = 119.84$, df = 2, and $P < 0.0005$ (using software, $P = 0.000$ to three decimal places). The largest contribution to the statistic comes from the youngest age group, which is cited more frequently than we would have expected under the hypothesis of no association. The other two age groups, which are cited less frequently than would be expected, also have large contributions.

**23.17:** STATE: Are all 12 astrological signs equally likely? PLAN: We test $H_0: p_1 = p_2 = ... = p_{12} = \frac{1}{12}$ vs. $H_a$: The 12 astrological sign birth probabilities are not equally likely. SOLVE: There are 1960 subjects in this sample. Under $H_0$, we expect $1960/12 = 163.33$ subjects per sign. Hence, all cells have expected counts greater than 5, and all cells have at least one observation. A chi-square test is appropriate. Hence, $\chi^2 = $

$\frac{(164 - 163.33)^2}{163.33} + \frac{(152 - 163.33)^2}{163.33} + ... + \frac{(177 - 163.33)^2}{163.33} = 16.09$. With df = 12 − 1 = 11, using Table D, $0.10 < P < 0.15$.

CONCLUDE: There is little evidence that some astrological signs are more likely in birth than others. That is, there is little to no support for a conclusion that astrological signs are not equally likely.

**23.19:** (b) $655/(655 + 916) = 655/1571 = 0.4169$.

**23.21:** (a) The expected cell count is $(1571)(1552)/4111 = 593.09$.

23.23: (a) df = $(r-1)(c-1) = (4-1)(2-1) = 3$.

23.25: (b) This is the hypothesis of association between "age" and "type of injury."

23.27: (b) We assume that the sample is an SRS, or essentially an SRS from all weightlifting injuries.

23.29: (a) The table below summarizes conditional distributions of opinion for each type of consumer. For example, there are $20 + 7 + 9 = 36$ buyers, so the proportion of buyers that think the quality of the recycled product is higher is $20/36 = 0.556$, or 55.6%.

Think the quality of recycled product is

|  | Higher | Same | Lower |
|---|---|---|---|
| Buyers | 55.6% | 19.4% | 25.0% |
| Nonbuyers | 29.9% | 25.8% | 44.3% |

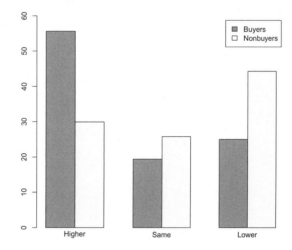

It seems that buyers of recycled products are more likely to feel that recycled products are of higher quality, while nonbuyers are more likely to feel that recycled products are of lower quality. (b) We test $H_0$: No association between "opinion of quality" and "buyer status" vs. $H_a$: There is some association between buyer status and opinion of quality. All expected cell counts are more than 5, so the guidelines for the chi-square test are satisfied. We have $\chi^2 = 7.64$, df = 2, and $0.01 < P < 0.025$. There is strong evidence of an association between buyer status and opinion of quality.

```
Minitab output
              Better    Same    Lower    Total
Buyers          20        7       9        36
              13.26     8.66    14.08

Nonbuyers       29       25      43        97
              35.74    23.34   37.92
Total           49       32      52       133
```

(c) We see that there is a relationship between opinion of quality and whether somebody buys the recycled product. However, it is impossible to determine whether (i) prior opinion on quality drives the decision to buy or not to buy; or (ii) perhaps quality of both types of products are excellent, and whichever product you happen to buy drives your opinion of that product… for example, if you buy a nonrecycled product and think it is of high quality, you might conclude (erroneously, perhaps) that the recycled product is of lower quality.

22.31: (a) The diagram is shown below. To perform the randomization, label the infants 01 to 77, and choose pairs of random digits. (b) See the Minitab output for the two-way table. We find $\chi^2 = 0.568$, df = 3, and $P = 0.904$. There is no reason to doubt that the randomization "worked."

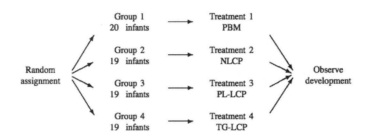

```
         Female    Male   Total
PBM        11        9      20
          10.91    9.09

NLCP       11        8      19
          10.36    8.64

PL-LCP     11        8      19
          10.36    8.64

TG-LCP      9       10      19
          10.36    8.64

Total      42       35      77

ChiSq =  0.001 +  0.001 +
         0.039 +  0.047 +
         0.039 +  0.047 +
         0.179 +  0.215 = 0.568
df = 3, p = 0.904
```

23.33: (a) We test $H_0: p_1 = p_2$ vs. $H_a: p_1 < p_2$. (b) The $z$ test must be used because the chi-square procedure measures evidence in support of evidence of any association, and is implicitly two-sided. We have $\hat{p}_1 = 0.3667$ and $\hat{p}_2 = 0.7333$. The pooled proportion is $\hat{p} = (11+22)/(30+30) = 0.55$, and the standard error is SE = 0.12845, so $z = -2.85$ and $P = 0.0022$. We have strong evidence that rats that can stop the shock (and therefore presumably have better attitudes) develop tumors less often than rats that cannot (and therefore are presumably depressed).

|         | Tumor | No tumor |
|---------|-------|----------|
| Group 1 | 11    | 19       |
| Group 2 | 22    | 8        |

23.35: STATE: Does sexual content of ads differ in magazines aimed at different audiences? PLAN: We test $H_0$: There is no relationship between sexual content of ads and magazine audience vs. $H_a$: There is some relationship between sexual content of ads and magazine audience. SOLVE: Examining the Minitab output, we see that conditions for use of the chi-square test are satisfied since all expected cell counts exceed 5. We obtain $\chi^2 = 80.874$ with df = 2, leading to $P < 0.0005$. CONCLUDE: Magazines aimed at women are much more likely to have sexual depictions of models than the other two types of magazines. Specifically, about 39% of ads in women's magazines show sexual depictions of models, compared with 21% and 17% of ads in general-audience and men's magazines, respectively. The two women's chi-squared terms account for over half of the total chi-square value.

23.37: We need cell counts, not just percents. If we had been given the number of travelers in each group — leisure and business — we could have estimated this.

22.39: In order to do a chi-square test, each subject can only be counted once.

23.41: (a) We test $H_0$: There is no relationship between degree held and service attendance vs. $H_a$: There is some relationship between degree held and service attendance. Examining the Minitab output, $\chi^2 = 14.19$ with df = 3, yielding $P$-value = 0.0027. There is strong evidence of an association between degree held and service attendance.

```
Minitab output
            HSchool    JColl    Bach    Grad    Total
Attend      400        62       146     76      684
            437.3      55.7     129.2   61.8

NoAttend    880        101      232     105     1318
            842.7      107.3    248.9   119.1

Total       1280       163      378     181     2002
```

ChiSq = 3.182 + 0.713 + 2.185 + 3.263 +
        1.651 + 0.370 + 1.147 + 1.669 = 14.19
df = 3, p = 0.0027

(b) The new table is shown below. We attain $\chi^2 = 0.73$ on df = 2. Hence, $P = 0.694$. In this table, we find no evidence of association between religious service attendance and degree held.

```
Minitab output
            JColl    Bach    Grad    Total
Attend      62       146     76      284
            64.1     148.7   71.2

NoAttend    101      232     105     438
            98.9     229.3   109.8

Total       163      378     181     722
```

ChiSq = 0.069 + 0.049 + 0.324 +
        0.045 + 0.032 + 0.210 = 0.729
df = 2, p = 0.694

(c) The new table is shown: We attain $\chi^2 = 13.40$ on df = 1. Hence, $P = 0.000$ to three places (it's actually 0.0002). There is overwhelming evidence of association between level of education (High school versus Beyond high school) and religious service attendance.

```
Minitab output
            HSchool    BeyondHS    Total
Attend       400        284         684
             437.3      246.7
NoAttend     880        438        1318
             842.7      475.3
Total       1280        722        2002

ChiSq = 3.182 + 5.640 +
        1.651 + 2.927 = 13.400
df = 1, p = 0.000
```

(d) In general, we find that people with degrees beyond high school attend service more often than expected, while people with high school degrees attend services less often than expected. Of those with high school degrees, 31.3% attended services, while the percentages are 38.0%, 38.6% and 42.0%, respectively, for people with Junior College, Bachelor's, and Graduate degrees.

23.43: STATE: Is there a relationship between race and parental opinion of schools? PLAN: We use a chi-square test to test $H_0$: There is no relationship between race and opinion about schools vs. $H_a$: There is some relationship between race and opinion about schools. SOLVE: All expected cell counts exceed 5, so use of a chi-square test is appropriate. We find that $\chi^2 = 22.426$ with df = 8 and $P = 0.004$. Nearly half of the total chi-square comes from the first two terms; most of the rest comes from the second and fifth rows. CONCLUDE: We have strong evidence of a relationship between race and opinion of schools. Specifically, according to the sample (as illustrated in the table), blacks are less likely and Hispanics are more likely to consider schools to be excellent, while Hispanics and whites differ in the percent considering schools good. Also, a higher percentage of blacks rated schools as "fair."

```
Minitab output
           Black    Hisp   White   Total
Exclnt       12      34      22      68
            22.70   22.70   22.59

Good         69      55      81     205
            68.45   68.45   68.11

Fair         75      61      60     196
            65.44   65.44   65.12

Poor         24      24      24      72
            24.04   24.04   23.92

DontKnow     22      28      14      64
            21.37   21.37   21.26

Total       202     202     201     605

ChiSq =   5.047 + 5.620 + 0.015 +
          0.004 + 2.642 + 2.441 +
          1.396 + 0.301 + 0.402 +
          0.000 + 0.000 + 0.000 +
          0.019 + 2.058 + 2.481 = 22.426
df = 8, p = 0.004
```

23.45: PLAN: We compare how detergent preferences vary by laundry habits. We use a chi-square test to test $H_0$: There is no relationship between laundry habits and preference vs. $H_a$: There is some relationship between laundry habits and preference. SOLVE: To compare people with different laundry habits, we compare the percent in each class who prefer the new product.

|  | Soft water, warm wash | Soft water, hot wash | Hard water, warm wash | Hard water, hot wash |
|---|---|---|---|---|
| Prefer new product | 54.3% | 51.8% | 61.8% | 58.3% |

The differences are not large, but the "hard water, warm wash" group is most likely to prefer the new detergent. With expected cell counts exceeding 5, a chi-square test is appropriate. We observe $\chi^2 = 2.058$ with df = 3, so that $P = 0.560$. CONCLUDE: The data provide no evidence to conclude that laundry habits and brand preference are related.

```
Minitab output
             S/W     S/H     H/W     H/H    Total
Standard      53      27      42      30     152
             49.81   24.05   47.23   30.92

New           63      29      68      42     202
             66.19   31.95   62.77   41.08

Total        116      56     110      72     354

ChiSq =   0.205 + 0.363 + 0.579 + 0.027
          0.154 + 0.273 + 0.436 + 0.020 = 2.058
df = 3, p = 0.560
```

**23.47: STATE:** How do the conditional distributions of political leaning, given education compare? **PLAN:** We compare the percentages leaning toward each party within each education group. **SOLVE:** The requested table is provided. At each education level, we compute the percentage leaning each party. For example, among Bachelor degree holders, 157/(157 + 154) = 50.5% lean Democrat, while the other 49.5% lean Republican.

|  | None | HS | JC | Bachelor | Grad | Total |
|---|---|---|---|---|---|---|
| Democrat | 142 | 500 | 81 | 157 | 103 | 983 |
|  | 68.9% | 60.8% | 55.1% | 50.5% | 63.6% |  |
| Republican | 64 | 323 | 66 | 154 | 59 | 666 |
|  | 31.1% | 39.2% | 44.9% | 49.5% | 36.4% |  |

**CONCLUDE:** At every education level, people leaning Democrat outweigh people leaning Republican. The difference is greatest at the "None" level of education, then decreases until the party support is nearly equal for Bachelor holders. Among graduate degree holders, Democrats strongly outnumber Republicans.

# Chapter 24: Inference for Regression

24.1: (a) A scatterplot of the data is provided, along with the least-squares regression line (students were not asked to add the regression line). We see that there is a strong linear relationship between wine intake and relative risk. From software, the correlation is $r = 0.985$.

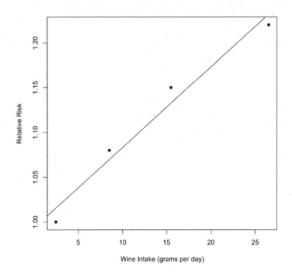

(b) The estimate of $\beta$ is $b = 0.0009$ (taken from output). We estimate that an increase in intake of 1 gram per day increases relative risk of breast cancer by 0.0009. The estimate of $\alpha$ is $a = 0.9931$ (see output). According to our estimate, wine intake of 0 grams per day is associated with a relative risk of breast cancer of 0.9931 (about 1).

```
Regression Analysis: RRisk versus Intake

The regression equation is RRisk = 0.993 + 0.009 Intake

Predictor     Coef      SE Coef        T           P
Constant    0.993094    0.017771    55.884      0.00032
Intake      0.009012    0.001112     8.102      0.01489

S = 0.01986    R-Sq = 97.0%    R-Sq(adj) = 95.6%
```

(c) The least-squares regression line is given by $\hat{y} = 0.9931 + 0.0009x$. The table below summarizes computed residuals, which sum to zero, as demonstrated. We also have $s^2 = 0.00079/2 = 0.000395$, which provides an estimate of $\sigma^2$. Hence, we estimate $\sigma$ by $s = \sqrt{0.000395} = 0.01987$, which agrees (up to rounding error) with "$S$" in the output above.

|      |      |        | Residual      |                 |
| x    | y    | $\hat{y}$ | $y - \hat{y}$ | $(y - \hat{y})^2$ |
|------|------|--------|---------------|-----------------|
| 2.5  | 1.00 | 1.0156 | −0.0156       | 0.00024         |
| 8.5  | 1.08 | 1.0697 | 0.0103        | 0.00011         |
| 15.5 | 1.15 | 1.1328 | 0.0172        | 0.00030         |
| 26.5 | 1.22 | 1.2319 | −0.0119       | 0.00014         |
|      |      |        | 0             | 0.00079         |

24.3: (a) A scatterplot of discharge by year is provided, along with the fitted regression line, which is requested in part (b). Discharge seems to be increasing over time, but there is also a lot of variation in this trend, and our impression is possibly influenced by the most recent years' data. From the output provided below, $r^2 = 0.225$, so the least-squares regression line explains 22.5% of the total observed variability in arctic discharge. (b) The regression line has been added to the scatterplot provided, and is given by $\hat{y} = -3690.08 + 2.80x$. We see from the output that $s = 111$.

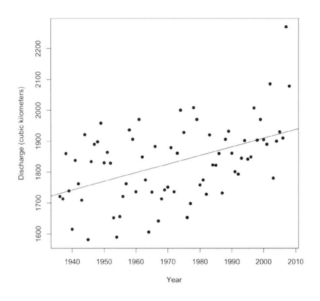

**Regression Analysis: Discarge versus Year**

The regression equation is Discharge = -3690.08 + 2.80Year

```
Predictor      Coef    SE Coef         T         P
Constant   -3690.078   3.29586     8.338    0.0000
Year           2.800    0.6168     4.539    0.0002

S = 111   R-Sq = 22.5%    R-Sq(adj) = 21.4%
```

24.5: Refer to the output provided with the solution to Exercise 24.3. We test $H_0 : \beta = 0$ vs. $H_a : \beta > 0$. We compute $t = \dfrac{b}{SE_b} = \dfrac{2.800}{0.6168} = 4.539$. Here, df $= n - 2 = 73 - 2 = 71$. In referring to Table C, we round df down to df $= 60$. Using Table C, we obtain $P < 0.0005$. Using software, we obtain $P = 0.0002$. There is strong evidence of an increase in arctic discharge over time.

*Solutions*

24.7: (a) For testing $H_0: \beta = 0$ vs. $H_a: \beta > 0$, we have $t = 8.104$ with df = 2. For the one-sided alternative suggested, we obtained $0.005 < P < 0.01$. This test is equivalent to testing $H_0$: Population correlation = 0 vs. $H_a$: Population correlation > 0. (b) Using software, $r = 0.985$. This can also be computed by referring to the Minitab output provided with Exercise 24.1, with $r = +\sqrt{r^2} = +\sqrt{0.97}$. Now, referring to Table E with $n = 4$, we find that $0.005 < P < 0.01$, just as in part (a). These tests are equivalent.

24.9: Referring to Table C, $t^* = 2.920$ (df = 4 – 2 = 2, with 90% confidence). Hence, a 90% confidence interval for $\beta$ is given by $0.009012 \pm 2.920(0.001112) = 0.009012 \pm 0.003247 = 0.00577$ to $0.01226$. With 90% confidence, the increase in relative risk of breast cancer associated with an increase in alcohol consumption by 1 gram per day is between 0.00577 and 0.01226.

24.11: Refer to the output provided in the solution to Exercise 24.3. We have $b = 2.800$ and $SE_b = 0.6168$. With 73 observations, df = 71. Using Table C, we look under the row corresponding to df = 60 (the nearest smaller value of df in the table). We obtain $t^* = 1.671$. Hence, a 90% confidence interval for $\beta$ is given by $2.8000 \pm 1.671(0.6168) = 2.8000 \pm 1.0307 = 1.7693$ to $3.8307$ cubic kilometers per year. With 90% confidence, the yearly increase in arctic discharge is between 1.7693 and 3.8307 cubic kilometers. This confidence interval excludes "0," so there is evidence arctic discharge is increasing over time.

24.13: (a) If $x^* = 0.60$, then our prediction for mean Aroc is $\hat{\mu} = 8.91 + 87.76(0.60) = 61.57$. (b) We have $SE_{\hat{\mu}} = 2.184$. For df = 29 – 2 = 27 and 95% confidence, we have $t^* = 2.052$. Hence, a 95% confidence interval for mean Aroc in people with 0.6 volume of the Brodmann area is given by $61.57 \pm 2.052(2.184) = 57.088$ to $66.052$.

24.15: (a) The residual plot provided does not suggest any deviation from a straight-line relationship between volume and Aroc score, although there are two residuals of larger magnitude present, both for Aroc scores slightly lower than 0.60.

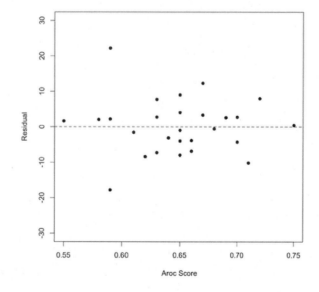

(b) A stemplot of residuals, provided below, does not suggest that the distribution of residuals departs strongly from Normality. There are two possible outliers, which agrees with the output provided by Minitab referenced in the problem statement.

```
-1 | 8
-1 | 0
-0 | 88777
-0 | 4443211
 0 | 0222333334
 0 | 889
 1 | 2
 1 |
 2 | 2
```

(c) It is reasonable to assume that observations are independent, since we have 29 different subjects, measured separately. (d) Referring to the residual plot of (b), it may be the case that spread is larger for smaller values of Aroc, but these happen to be the two outliers. It is difficult to make a definitive argument either way.

24.17: (a) With a positive association, $r = +\sqrt{r^2} = +\sqrt{0.623} = 0.789$.

24.19: (a) This is a one-sided alternative, because we wonder if larger appraisal values are associated with larger selling prices.

24.21: (c) $s = 235.41$.

24.23: (c) With 26 degrees of freedom, $t^* = 2.056$, so the margin of error is $2.056(0.1938) = 0.3985$.

24.25: (a) Scientists estimate that each additional 1% increase in the percentage of Bt cotton plants results in an increase of 6.81 mirid bugs per 100 plants. (b) The regression model explains 90% of the variability in mirid bug density. That is, knowledge of the proportion of Bt cotton plants explains most of the variation in mired bug density. (c) Recall that the test $H_0: \beta = 0$ vs. $H_a: \beta > 0$ is exactly the same as the test $H_0$: Population correlation $= 0$ vs. $H_a$: Population correlation $> 0$. As $P < 0.0001$, there is strong evidence of a positive linear relationship between the proportion of Bt cotton plants and the density of mirid bugs. (d) We may conclude that denser mirid bug populations are associated with larger proportions of Bt cotton plants. However, it seems plausible that a reduced use of pesticides (an indirect cause), rather than more Bt cotton plants (a direct cause) is the reason for this increase.

24.27: For 90% intervals with df = 10, use $t^* = 1.812$. (a) Use the estimated slope and standard error given in Figure 24.13. The confidence interval for $\beta$ is $b \pm t^* SE_b = 274.78 \pm 1.812(88.18) = 274.78 \pm 159.78 = 115.0$ to $434.6$ fps/inch. (b) This is the "90% CI" given in Figure 24.13: 176.2 to 239.4 fps. To confirm this, we can use the given values of $\hat{y} = 207.8$ and $SE_{\hat{\mu}} = 17.4$, labeled "Fit" and "SE Fit" in the output:
$\hat{y} \pm t^* SE_{\hat{\mu}} = 207.8 \pm 1.812(17.4) = 176.3$ to $239.3$ fps, which agrees with the output up to rounding error.

**24.29:** (a) The stemplot has split stems. There is little evidence of non-Normality in the residuals, and there don't appear to be any strong outliers. (b) The scatterplot confirms the comments made in the text: there is no clear pattern, but the spread about the line may be slightly greater when $x$ is large. (c) Presumably, close inspection of a manatee's corpse will reveal non-subtle clues when cause of death is from collision with a boat rotor. Hence, it seems reasonable that the number of kills listed in the table are mostly not caused by pollution.

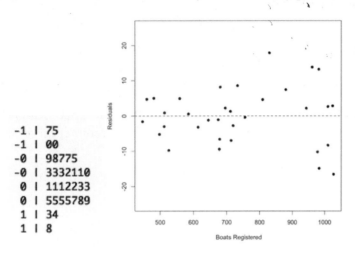

```
-1 | 75
-1 | 00
-0 | 98775
-0 | 3332110
 0 | 1112233
 0 | 5555789
 1 | 34
 1 | 8
```

**24.31:** (a) This is a confidence interval for $\beta$. With df = 31, using the table (and rounding degrees of freedom down to 30) we have $t^* = 2.042$ so a 95% confidence interval for $\beta$ is $b \pm t^* SE_b = 0.129232 \pm 2.042(0.00752) = 0.129232 \pm 0.01536 = 0.11387$ to $0.14459$ additional killed manatees per 1000 additional boats. (b) With 1,050,000 boats, we predict $\hat{y} = -43.17195 + 0.129232(1050) = 92.5217$ killed manatees, which agrees with the output in Figure 24.14 under "Fit." We need the prediction interval, since we are forecasting the number of manatees killed for a single year. According to the output provided, a 95% prediction interval for the number of killed manatees if 1,050,000 boats are registered is 75.18 to 109.87 kills.

**24.33:** See the output corresponding to a regression analysis, below. (a) We test $H_0$: Population correlation = 0 against $H_a$: Population correlation is positive. We see that $t = 3.88$ with df = 27 − 2 = 25. Hence, $P < 0.0005$. There is very strong evidence of a positive correlation between Gray's forecasted number of storms and the number of storms that actually occur. (b) The output provided provides the corresponding confidence interval for the mean number of storms in years for which Gray predicts 16 stores. Here, $\hat{\mu} = 1.022 + 0.9696(16) = 16.535$, and $SE_{\hat{\mu}} = 1.3070$ (obtained from output). With df = 25, $t^* = 2.060$ and the 95% confidence interval is given by $16.535 \pm 2.060(1.3070) = 13.843$ to $19.227$ storms.

**Regression Analysis: Observed versus Forecast**

The regression equation is
Observed = 1.02 + 0.970 Forecast
Predictor    Coef     SE Coef    T      P
Constant     1.022    3.008      0.34   0.737
Forecast     0.9696   0.2501     3.88   0.001
S = 3.79361   R-Sq = 37.5%   R-Sq(adj) = 35.0%
Fit       SE Fit   95% CI             95% PI
16.535    1.307    (13.843, 19.226)   (8.271, 24.799)

24.35: The stemplot is provided, where residuals are rounded to the nearest tenth. The plot suggests that the residuals do not follow a Normal distribution. Specifically, there are a number of rather extreme outliers. This makes regression inference and interval procedures unreliable.

```
 -7 | 5
 -6 |
 -5 |
 -4 | 7
 -3 | 75
 -2 | 88776
 -1 | 8
 -0 | 876
  0 | 3345
  1 | 334
  2 | 333
  3 | 23
  4 |
  5 |
  6 | 3
  7 |
  8 |
  9 |
 10 |
 11 | 4
```

24.37: (a) Shown is the scatterplot with two (nearly identical) regression lines: One using all points, and tone with the outlier omitted. Minitab output for both regressions is provided below. (b) The correlation for all points is $r = 0.8486$. For testing the slope, $t = 6.00$, for which $P < 0.001$. (c) Without the outlier, $r = 0.7014$, the test statistic for the slope is $t = 3.55$, and $P = 0.004$. In both cases there is strong evidence of a linear relationship between neural loss aversion and behavioral loss aversion. However, omitting the outlier weakens this evidence somewhat.

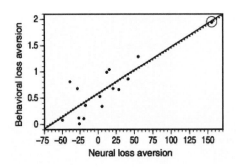

**Minitab output: Regression for all points**

```
The regression equation is Behave = 0.585 + 0.00879 Neural

Predictor      Coef      Stdev    t-ratio       p
Constant    0.58496    0.07093       8.25   0.000
Neural     0.008794   0.001465       6.00   0.000

s = 0.2797     R-sq = 72.0%    R-sq(adj) = 70.0%
```

**Minitab output: Regression with outlier omitted**

```
The regression equation is Behave = 0.586 + 0.00891 Neural

Predictor      Coef      Stdev    t-ratio       p
Constant    0.58581    0.07506       7.80   0.000
Neural     0.008909   0.002510       3.55   0.004

s = 0.2903     R-sq = 49.2%    R-sq(adj) = 45.3%
```

24.39: The distribution is skewed right, but the sample is large, so $t$ procedures should be safe. We find $\bar{x} = 0.2781$ g/m$^2$ and $s = 0.1803$ g/m$^2$. Table C gives $t^* = 1.984$ for df = 100 (rounded down from 115). Hence, the 95% confidence interval for $\mu$ is 0.2449 to 0.3113 g/m$^2$.

```
 0 | 0067778999
 1 | 000011112222333345555556667777788889
 2 | 00000111122333444666667788889
 3 | 00000111222333456667788999
 4 | 01456667
 5 | 3589
 6 | 04
 7 |
 8 | 29
 9 | 0
10 | 5
```

24.41: PLAN: We examine the relationship between pine cone abundance and squirrel reproduction using a scatterplot and regression. SOLVE: Regression gives predicted offspring $\hat{y} = 1.4146 + 0.4399x$. The slope is significantly different from zero ($t = 4.33$, $P = 0.001$). To assess the evidence that more cones leads to more offspring, we should use the one-sided alternative, $H_a : \beta > 0$, for which $P$ is half as large (so $P < 0.001$). The conditions for inference seem to be satisfied. The year-to-year observations should be reasonably independent, a stemplot of the residuals against cone abundance does not show any obvious non-normality. The residual plot shows no departures from a linear pattern. One might also choose to find a confidence interval for $\beta$: With df = 14, $t^* = 2.145$ for 95% confidence. Hence, a 95% confidence interval for $\beta$ is $0.4399 \pm 2.145(0.1016) = 0.2220$ to 0.6578 offspring per cone. CONCLUDE: We have strong evidence of a positive linear relationship between cone abundance and squirrel offspring. Specifically, we are 95% confidence that for each additional one-unit change in the cone index, the mean number of offspring increases by between 0.22 and 0.66 offspring per female.

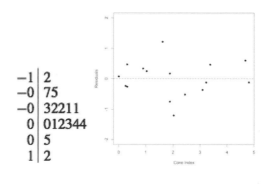

```
-1 | 2
-0 | 75
-0 | 32211
 0 | 012344
 0 | 5
 1 | 2
```

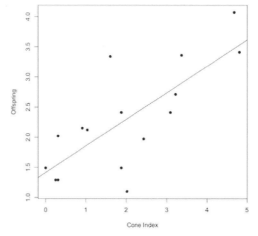

**Minitab output: Regression of offspring on cones**

```
The regression equation is Offspr = 1.41 + 0.440 Cones

Predictor      Coef      Stdev     t-ratio       p
Constant     1.4146     0.2517       5.62    0.000
Cones        0.4399     0.1016       4.33    0.001

s = 0.6003    R-sq = 57.2%    R-sq(adj) = 54.2%
```

24.43: PLAN: We will examine the relationship between beaver stumps and beetle larvae using a scatterplot and regression. We specifically wish to test for a positive slope $\beta$, and find a confidence interval for $\beta$. SOLVE: The scatterplot shows a positive linear association; the regression line is $\hat{y} = -1.286 + 11.894x$. A stemplot of the residuals does not suggest non-normality of the residuals, the residual plot does not suggest non-linearity, and the problem description makes clear that observations are independent. To test $H_0 : \beta = 0$ vs. $H_a : \beta > 0$, the test statistic is $t = 10.47$ (df=21), for which Table C provides a one-sided $P$-value, $P < 0.0005$. For df = 21, $t^* = 2.080$ for 95% confidence, so with $b$ and $SE_b$ as given by Minitab, we are 95% confident that $\beta$ is between 9.531 and 14.257. CONCLUDE: We have strong evidence that beetle larvae counts increase with beaver stump counts. Specifically, we are 95% confident that each additional stump is (on the average) accompanied by between 9.5 and 14.3 additional larvae clusters.

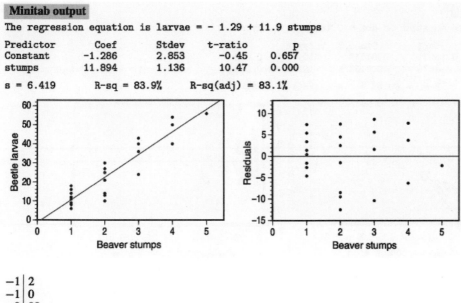

```
-1 | 2
-1 | 0
-0 | 98
-0 | 6
-0 | 4
-0 | 22
-0 | 11
 0 | 011
 0 | 233
 0 | 455
 0 | 777
 0 | 8
```

24.45: PLAN: Using a scatterplot and regression, we examine how well phytopigment concentration explains DNA concentration. SOLVE: The scatterplot shows a fairly strong linear positive association; the regression equation is $\hat{y} = 0.1523 + 8.1676x$. A stemplot of the residuals looks reasonably Normal, but the scatterplot suggests that the spread about the line is greater when phytopigment concentration is greater. This may make regression inference unreliable, but we will proceed. Finally, observations are independent, from the context of the problem. The slope is significantly different from 0 ($t = 13.25$, df = 114, $P < 0.001$). We might also construct a 95% confidence interval for $\beta$: $8.1676 \pm 1.984(0.6163) = 6.95$ to $9.39$. CONCLUDE: The significant linear relationship between phytopigment and DNA concentrations is consistent with the belief that organic matter settling is a primary source of DNA. Starting from a measurement of phytopigment concentration, we could give a fairly accurate prediction of DNA concentration, as the relationship explains about $r^2 = 60.6\%$ of the variation in DNA concentration. We are 95% confident that each additional unit increase in phytopigment concentration increases DNA concentration by between 6.95 and 9.39 units (on the average).

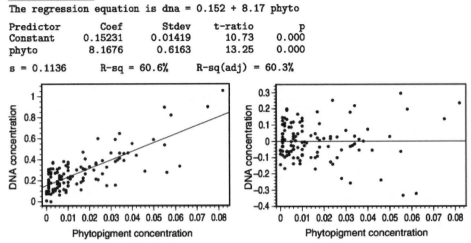

```
-3 | 32
-2 | 5
-2 | 42
-1 | 76
-1 | 443321000
-0 | 9999888887666666655555
-0 | 444333333222221100000
 0 | 000011111112222233334444
 0 | 66678899
 1 | 0011112233444
 1 | 678999
 2 | 13
 2 | 59
```

24.47: (a) The mean is $\bar{x} = -0.00333$, and the standard deviation is $s = 1.0233$. For a standardized set of values, we expect the mean and standard deviation to be (up to rounding error) 0 and 1, respectively. (b) The stemplot does not look particularly symmetric, but it is not strikingly non-Normal for such a small sample. (c) The probability that a standard Normal variable is as extreme as this is about 0.0272.

```
-2 | 2
-1 |
-1 | 4
-0 |
-0 | 32
 0 | 01122
 0 | 7
 1 | 0
 1 | 5
```

24.49: For df = 14 and a 95% confidence interval, we use $t^* = 2.145$, so the interval is $-0.01270 \pm 2.145(0.01264) = -0.0398$ to $0.0144$. This interval does contain 0.

# Chapter 25: One-Way Analysis of Variance: Comparing Several Means

25.1: (a) We wish to test whether the average status measure in the four groups (men expressing anger, men expressing sadness, women expressing anger, women expressing sadness) differ. That is, we test $H_0: \mu_A = \mu_B = \mu_C = \mu_D$ vs. $H_a$: Not all means agree. (b) Referring to Figure 25.2, comparing Groups A and C, we see that the mean status for men expressing anger is about 6.3, while the mean status for women expressing anger is about 4. Hence, with both groups expressing anger, men receive higher mean status scores than women, and the mean difference is about 2.3. Notice that comparing Groups B and D, we see that women expressing sadness receive higher status scores than men expressing sadness, but the difference is relatively small.

25.3: (a) The stemplots appear to suggest that logging reduces the number of trees per plot and that recovery is slow (the 1-year-after and 8-years-after stemplots are similar). (b) The means lead one to the same conclusion as in (a): The first mean is much larger than the other two. (c) In testing $H_0: \mu_1 = \mu_2 = \mu_3$ vs. $H_a$: Not all means are the same, we find that $F = 11.43$ with df = 2 and 30, which has $P = 0.000205$, so we conclude that these differences are significant: the number of trees per plot really is lower in logged areas.

```
Never logged  1 year ago   8 years ago
0|            0|2          0|4
0|            0|9          0|
1|            1|2244       1|22
1|699         1|57789      1|5889
2|0124        2|0          2|22
2|7789        2|           2|
3|3           3|           3|
```

25.5: (a) By moving the middle mean to the same level as the other two, it is possible to reduce $F$ to about 0.02, which has a $P$-value very close to the left end of the scale (near 1). (b) By moving any mean about 1 centimeter up or down (or any two means about 0.5 cm in opposite directions), the value of $F$ increases (and $P$ decreases) until it moves to the right end of the scale.

25.7: (a) We have $s_1^2 = 25.6591$, $s_2^2 = 24.8106$, and $s_3^2 = 33.1944$. Hence, $s_1 = 5.065$, $s_2 = 4.981$ and $s_3 = 5.761$. The ratio of largest to smallest standard deviation is $5.761/4.981 = 1.16$, which is less than 2. Conditions are satisfied. (b) The three standard deviations are $s_L = 16.61$, $s_M = 17.42$ and $s_C = 17.13$. Hence, the ratio of largest to smallest standard deviation is $17.42/16.61 = 1.05$, which is less than 2. Conditions are satisfied.

25.9: STATE: How does the presence of nitrogen, phosphorus, or both affect the development of new leaves in bromeliads? PLAN: Examine the data to compare the effect of the treatments and check that we can safely use ANOVA. If the data allow ANOVA, assess the significance of observed differences in mean numbers of new leaves. SOLVE: Side-by-side stemplots shows some irregularity, but no outliers or strong skewness.

The Minitab ANOVA output below shows that the group standard deviations easily satisfy our rule of thumb (2.059/1.302 = 1.58 < 2). The differences among the groups were significant at $\alpha = 0.05$: $F = 3.44$, df=3 and 27, $P = 0.031$. CONCLUDE: Nitrogen had a positive effect, the phosphorus and control groups were similar, and the plants that got both nutrients fell between the others.

**25.11:** (a) $I = 3$ and $N = 96$, so df = 2 and 93. (b) $I = 3$ and $N = 90$, so df = 2 and 87.

**25.13:** (a) No sample standard deviation is larger than twice any other. Specifically, the ratio of largest to smallest standard deviation is 2.25/1.61 = 1.40, which is less than 2. Conditions are safe for use of ANOVA. (b) Calculations are provided:

$$\bar{x} = \frac{17 \times 6.47 + 17 \times 3.75 + 17 \times 4.05 + 17 \times 5.02}{68} = 4.8225$$

$$\text{MSG} = \frac{17(6.47 - 4.8225)^2 + 17(3.75 - 4.8225)^2 + 17(4.05 - 4.8225)^2 + 17(5.02 - 4.8225)^2}{4 - 1} = 25.502$$

$$\text{MSE} = \frac{(17-1)2.25^2 + (17-1)1.77^2 (17-1)1.61^2 + (17-1)1.80^2}{68 - 4} = 3.507$$

$$F = \frac{\text{MSG}}{\text{MSE}} = \frac{25.502}{3.507} = 7.272$$

(c) We have df = 4 − 1 = 3 and 68 − 4 = 64, so we refer to the $F$ distribution with 3 and 64 degrees of freedom. Citing the output provided below, the P-value is 0.000 rounded to three decimal places. In fact, $P = 0.0003$ (obtained using software). There is strong evidence that the mean status scores between the four groups studied are not equal—a conclusion consistent with the solution to Exercise 25.1.

```
Minitab output

Source    DF    SS        MS       F        p
Factor    3     76.504    25.502   7.272    0.000
Error     64    224.440   3.507
Total     67    300.944
```

**25.15:** (c) the means of several populations.

**25.17:** (b) $I - 1 = 3 - 1 = 2$, and $N - I = 9 - 3 = 6$.

**25.19:** (c) Since MSG = 22,598/(3 − 1) = 22,598/2 = 11,299, $F$ = MSG/MSE = 11,299/1600 = 7.06.

25.21: (c) The largest standard deviation is 62.02, and the smallest is 20.07. Hence, the largest standard deviation is more than twice the smallest.

25.23: (c) We do not have three independent samples from three populations.

25.25: The populations are college students that might view the advertisement with art image, college students that might view the advertisement with a nonart image, and college students that might view the advertisement with no image. The response variable is student evaluation of the advertisement on the 1–7 scale. We test the hypothesis $H_0 : \mu_1 = \mu_2 = \mu_3$ (all three groups have equal mean advertisement evaluation) vs. $H_a$: Not all means are equal. There are $I = 3$ populations; the samples sizes are $n_1 = n_2 = n_3 = 39$, so there are $N = 39 + 39 + 39 = 117$ individuals in the total sample. There are then $I - 1 = 3 - 1 = 2$ and $N - I = 117 - 3 = 114$ df.

25.27: The response variable is hemoglobin A1c level. We have $I = 4$ populations; a control (sedentary) population, an aerobic exercise population, a resistance training population, and a combined aerobic and resistance training population. We test hypothesis $H_0 : \mu_1 = \mu_2 = \mu_3 = \mu_4$ (all four groups have equal mean hemoglobin A1c levels) vs. $H_a$: Not all means are equal. Sample sizes are $n_1 = 41$, $n_2 = 73$, $n_3 = 72$, and $n_4 = 76$. Our total sample size is $N = 41 + 73 + 72 + 76 = 262$. We have $I - 1 = 4 - 1 = 3$ and $N - I = 262 - 4 = 258$ df.

25.29: (a) The graph suggests that emissions rise when a plant is attacked because the mean control emission rate is half the smallest of the other rates. (b) The null hypothesis is "all groups have the same mean emission rate." The alternative is "at least one group has a different mean emission rate." (c) The most important piece of additional information would be whether the data are sufficiently close to Normally distributed. (From the description, it seems reasonably safe to assume that these are more or less random samples.) (d) The SEM equals $s/\sqrt{8}$, so we can find the standard deviations by multiplying by $\sqrt{8}$; they are 16.77, 24.75, 18.78, and 24.38. However, this factor of $\sqrt{8}$ would cancel out in the process of finding the ratio of the largest and smallest standard deviations, so we can simply find this ratio directly from the SEMs: $\dfrac{8.75}{5.93} = \dfrac{24.75}{16.77} = 1.48$, which satisfies our rule of thumb.

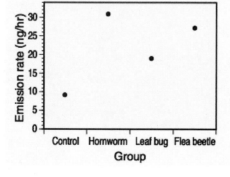

25.31: (a) The stemplots are provided, and means and standard deviations are in the Minitab output. The means suggest that extra water in the spring has the greatest effect on biomass, with a lesser effect from added water in the winter. ANOVA is risky with these data; the standard deviation ratio is nearly 3, and the winter and spring distributions may have skewness or outliers (although it is difficult to judge with such small samples). (b) We wish to test whether the mean biomass from any group differs from the others: $H_0: \mu_w = \mu_s = \mu_c$ vs. $H_a$: At least one mean is different. (c) ANOVA gives a statistically significant result ($F = 27.52$, df 2 and 15, $P < 0.0005$), but as noted in (a), the conditions for ANOVA are not satisfied. Based on the stemplots and the means, however, we should still be safe is concluding that added water increases biomass.

```
Minitab output
Source       DF      SS         MS       F       p
Treatment     2    97583      48791    27.52   0.000
Error        15    26593       1773
Total        17   124176
                                   Individual 95% CIs For Mean
                                   Based on Pooled StDev
Level     N     Mean    StDev    ------+---------+---------+---------+
winter    6    205.17   58.77                (----*-----)
spring    6    315.39   37.34                              (----*----)
control   6    136.65   21.69    (-----*----)
                                 ------+---------+---------+---------+
Pooled StDev =    42.11             140       210       280       350
```

Winter stemplot:
```
1|
1|
1|4
1|67
1|8
2|
2|
2|6
2|9
3|
3|
3|
3|
3|
```

Spring stemplot:
```
1|
1|
1|
1|
1|
2|
2|
2|
2|
2|889
3|1
3|2
3|
3|
3|8
```

Control stemplot:
```
1|11
1|2
1|44
1|7
1|
2|
2|
2|
2|
2|
3|
3|
3|
3|
3|
```

25.33: (a) The design, with four treatments, is shown. (b) PLAN: We compare the mean lightness of the four methods using a plot of the means and ANOVA. SOLVE: ANOVA should be safe: It is reasonable to view the samples as SRSs from the four populations, the distributions do not show drastic deviations from Normality, and the standard deviations satisfy our rule of thumb ($0.392/0.250 = 1.568$). The Minitab output below includes a table of the means, and a display that is equivalent to a plot of the means. The means show rather small differences in lightness score; Method C is lightest and Method B is darkest. The differences in mean lightness are nonetheless highly significant ($F = 22.77$, $P < 0.001$). CONCLUDE: The manufacturer will prefer Method B. Whether these differences are large enough to be important in practice requires more information about the scale of lightness scores.

```
Method A    Method B    Method C    Method D
40|         40|89       40|         40|
41|1        41|         41|         41|
41|2        41|2233     41|         41|
41|44       41|5        41|         41|
41|7        41|6        41|         41|667
41|8        41|         41|         41|99
42|0        42|         42|         42|0
42|2        42|         42|223      42|23
42|         42|         42|445      42|
42|         42|         42|6        42|
42|         42|         42|         42|
43|         43|         43|1        43|
```

Design diagram: Random assignment splits into Group 1 (8 pieces) → Treatment 1 Method A; Group 2 (8 pieces) → Treatment 2 Method B; Group 3 (8 pieces) → Treatment 3 Method C; Group 4 (8 pieces) → Treatment 4 Method D; then Compare lightness.

# Solutions

```
Minitab output
Source      DF       SS        MS        F        p
Method       3    6.2815    2.0938    22.77    0.000
Error       28    2.5752    0.0920
Total       31    8.8567
                                       Individual 95% CIs For Mean
                                       Based on Pooled StDev
Level        N     Mean      StDev   ---------+---------+---------+--------
A            8    41.649    0.392              (---*---)
B            8    41.283    0.255    (----*---)
C            8    42.493    0.294                              (----*---)
D            8    41.950    0.250                    (---*---)
                                     ---------+---------+---------+--------
Pooled StDev =    0.303                     41.50     42.00     42.50
```

**25.35:** STATE: Are the mean tip percentages constant for all types of weather forecasts (no forecast, good forecast, bad forecast)? PLAN: We will carry out an ANOVA test for the equality of means. SOLVE: First, we see that the ratio of largest standard deviation to smallest standard deviation is 2.388/1.959 = 1.219, which is less than 2. Histograms of the samples are provided. There is some evidence of non-Normality, and perhaps one outlier in the "No Weather Report" group. We proceed, as the samples are reasonably large. From the output, we have $F = 20.679$ on $3 - 1 = 2$ and $60 - 3 = 57$ df. Hence, P = 0.000. CONCLUDE: There is overwhelming evidence that the mean tip percentages are not the same for all three groups. Examination of the summary statistics and the histograms provided suggests that while mean tip for the Bad report group is similar to that of the No report group, the mean tip for Good weather report is higher.

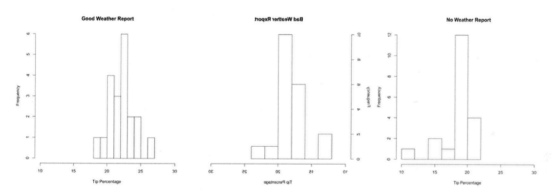

```
Minitab output
Source      DF        SS        MS        F        p
Report       2    192.224    96.112    20.679   0.000
Error       57    264.922     4.648
Total       59    457.146

            Mean     SD      n
Good       22.22    1.959    20
Bad        18.18    2.098    20
None       18.73    2.388    20
```

**25.37:** First, we note that the mean angle for untreated fabric is 79 degrees, showing much less wrinkle resistance than any of the treated fabrics. ANOVA on four groups gives $F = 153.76$ and $P < 0.001$. A comparison of wrinkle recovery angle for the three durable press treatments is more interesting. The ANOVA results are shown in the Minitab output below. Hylite LF, which in Exercise 25.36 was seen to have the lowest breaking strength, has the highest wrinkle resistance. There is almost no difference between the means for two versions of Permafresh, even though we saw in Exercise 25.36 that Permafresh 55 appears to be stronger than Permafresh 48. The ANOVA $F$-test cannot be trusted because the standard deviations violate our rule of thumb: 10.16/1.92 = 5.29. This is much larger than 2. In particular, Permafresh 48 shows much more variability from piece to piece than either of the other treatments. Large variability in performance is a serious defect in a commercial product, so it appears that Permafresh 48 is unsuited for use on these grounds. The data are very helpful to a maker of durable press fabrics despite the fact that the formal test is not valid.

```
Minitab output
Source      DF      SS       MS       F       p
treat        2     253.2    126.6    3.39    0.068
Error       12     448.4     37.4
Total       14     701.6
                                  Individual 95% CIs For Mean
                                  Based on Pooled StDev
Level        N      Mean    StDev  ----+---------+---------+---------+----
P55          5    134.80     1.92     (---------*---------)
P48          5    134.20    10.16  (---------*---------)
HY           5    143.20     2.28                    (---------*---------)
                                  ----+---------+---------+---------+----
Pooled StDev =      6.11           132.0     138.0     144.0
```

25.39: (a) There is a slight increase in growth when water is added in the wet season, but a much greater increase when it is added during the dry season. (b) The means differ significantly during the first three years. (c) The year 2005 is the only one for which the winter biomass was higher than the spring biomass.

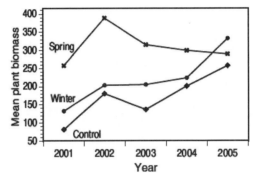

25.41: In addition to a high standard deviation ratio (117.18/35.57 = 3.29), the spring biomass distribution has a high outlier.